Pocket Guide...

Preventing Process Plant Materials Mix-ups

Gulf Publishing Company
Houston, Texas

Pocket Guide to Preventing Process Plant Materials Mix-ups

Pocket Guide to
Preventing Process Plant Materials Mix-ups

Gulf Publishing Company
Book Division
P.O. Box 2608 □ Houston, Texas 77252-2608

10 9 8 7 6 5 4 3 2 1

Library of Congress Cataloging-in-Publication Data

Moniz, Bert.
Pocket guide to preventing process plant materials mix-ups / by Bert Moniz.
p. cm.
ISBN 0-88415-344-4 (alk. paper)
1. Materials management—Handbooks, manuals, etc. 2. Industrial procurement—Handbooks, manuals, etc. I. Title: Preventing process plant materials mix-ups. II. Title.
TS161.M66 2000
658.7—dc21 00-025085

Printed in the United States of America.
Printed on acid-free paper (∞).

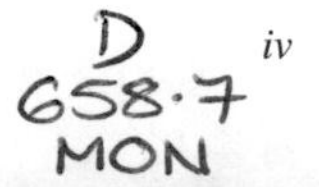

Dedication

To Greg Korbin and Phil Peters—technically swift, yet disarmingly practical. And to Loretta, Valerie and Carol—what better reasons to believe in the future.

Contents

Preface

Materials mix-ups are serious problems in process plants, and occur when the wrong component, part, or material of construction is substituted during the fabrication or assembly of equipment, delivered to the site, or installed in process. Materials mix-ups have led to catastrophic incidents. Additionally, premature failures of improperly substituted parts from materials mix-ups have led to significant production losses and costly equipment rework.

Materials mix-ups may occur at any stage in the product supply chain. Strict control procedures must be implemented, at the interfaces where information or product is transferred from one organization to another.

This pocket guide describes effective methods of specifying, procuring, receiving, and verifying critical materials. It describes how materials are tested and identified, and explains the differences between the various production methods for materials. The object is to provide all people in the supply chain with the tools to prevent costly materials mix-ups.

The following examples of materials mix-ups illustrate the diversity of this problem:

Problem	Consequence	Solution
Bearings: Undersized replica parts substituted for original equipment size on 200 hp blower	Excessive vibration caused premature failure of blower with shutdown to replace bearings	Specify replica parts accurately and know if the are applicable
Fasteners: Vendor-supplied wrong grade (SAE J429) bonnet bolts in a valve	Valve bolts failed catastrophically releasing sulfuric acid resulting in personal injury	Ensure suppliers of components that contain embedded critical materials have adequate control of their supply chains
Flanges: Fraudulent, inadequate flanges from other countries escaped detection and were used on domestically produced equipment and piping systems	Excessive search and removal from installed equipment and piping systems where catastrophic failure might have resulted	Ensure all suppliers pass audit and have adequate quality control procedures for their supply chains
Welding Filler Metal: Wrong filler metal with poorer corrosion resistance (317L stainless steel) was used in alloy C-276 piping	Product leakage at welds required shutdown and replacement of piping	Check certification of weld filler metals used for field fabrication. For critical applications conduct chemical analysis on sample
Seal: Substitution of cheap compounding ingredients	High volume swell and loss of sealing capability	Accurately specify all major ingredients in rubber and plastic components

This pocket guide is a detailed reference for the following people or job functions: engineers, designers, materials clerks, planners, stores personnel, buyers, suppliers of maintenance, repair and operations (MRO) materials, equipment fabricators, valve and component rebuilders, mechanics, operators performing TPM (total productive manufacturing), process safety managers, maintenance supervisors, and reliability engineers.

Success means improved safety, less downtime, better profitability and greater job security!

Bert Moniz

1

Quality Control Systems

Summary: This chapter describes a four-component program of a program to avoid materials mix-ups on your site: 1) effective identification of materials and parts in your procurement database, 2) selection of competent suppliers, 3) organization of efficient receiving inspections, and 4) implementing field audit before parts are used. The regulatory background that drives us to avoid materials mix-ups is also described.

Process safety management is the primary requirement that drives process plants and refineries to establish quality control programs for incoming materials and spare parts. OSHA 29 CFR 1910.119 has defined regulations for process safety critical equipment and systems that include such requirements. Other reasons for quality control programs may be equally important, for example when failure has a significant impact on capability to make product or leads to excessive maintenance costs.

OSHA REQUIREMENTS FOR CRITICAL MATERIALS

OSHA 1910.119 describes process safety management regulations for highly hazardous chemicals, explosives and blasting agents, with the prime purpose of ". . . preventing or minimizing the consequences of catastrophic releases of toxic, reactive, flammable, or explosive chemicals." The OSHA regulation identifies specific chemicals and the quantities that cause them to be defined as hazardous. They include toxic and reactive compounds, flammable liquids and gases, and explosives and pyrotechnics.

An integral component of the OSHA regulations concerns quality assurance for the following types of equipment: pressure vessels and storage tanks; piping systems; relief and vent systems and devices; emergency shutdown systems; controls and pumps. The regulations state the following:

- *Appropriate checks and inspections shall be performed to assure that equipment is installed properly and consistent with design specifications and the manufacturer's instructions*

- *The employer shall assure that maintenance materials, spare parts and equipment are suitable for the process application for which they will be used*

CRITICAL MATERIALS SUPPLY CHAIN

Specific job functions are responsible for maintaining control at each stage of the critical materials supply chain. See Figure 1-1.

For example, consider what could go wrong with a fastener at the various stages of the supply chain:

Stage 1: Wrong material for job, e.g., carbon steel vs. stainless

Stage 2: Wrong manufacturing technique, e.g., rolled threads vs. machined threads

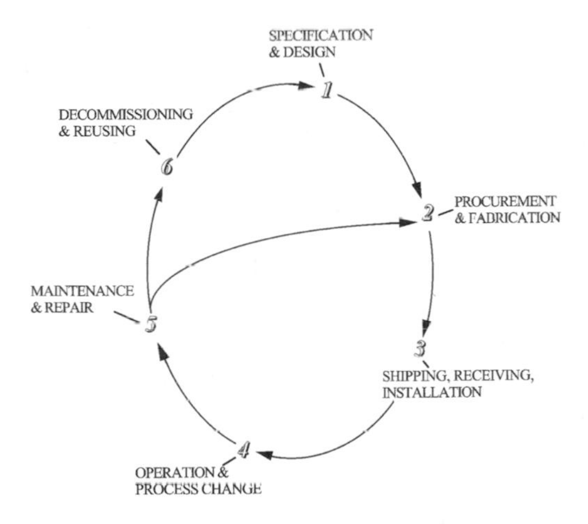

Figure 1-1. Individual job functions come into play at every step of the materials lifecycle.

Stage 3: Wrong item selected, e.g., mix-up in bins or boxes

Stage 4: Process change without recognizing need for new fastener type, e.g., a flammable liquid requires stronger fastener for spiral wound gaskets

Stage 5: Poorly described replacement parts in database, e.g., no approved supplier indicated

Stage 6: No procedure for reuse, e.g., no practice for decontamination, cleaning, or inspection

QUALITY CONTROL SYSTEM

A quality control system to prevent costly materials mix-ups consists of the following parts: 1) effective materials description in procurement database, 2) selection of competent suppliers, 3) receiving inspections performed at a mutually agreed level, and 4) field verification just before the materials are put to use. See Figure 1-2.

Effective Materials and Spare Parts Description in Procurement Database. Improper stores or procurement descriptions allow inadequate parts, materials or components to enter the plant, increasing the potential for process safety incidents (if under specified) or cost overruns (if over specified). Personnel entering parts descriptions in electronic or paper ordering systems must be trained to ensure the accuracy of their work. See Chapter 2 under Database Discipline.

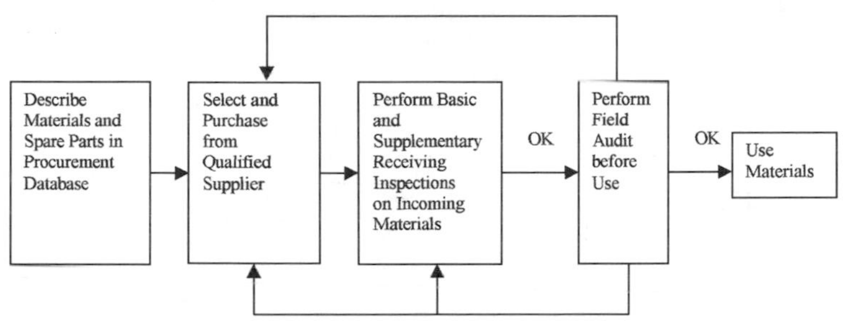

Figure 1-2. Four main steps are required in a quality control system to prevent materials mix-up.

The procurement system database must have an adequate number of fields for accurate part description, such as specification number and dimensions. There should also be a field to indicate whether the part is process safety critical, production critical, or maintenance critical. It allows the part to be flagged for special inspections or other considerations. Many electronic procurement systems are populated with inadequate or incorrect information. Job one is to clean them up. See Figure 1-3.

Select and Purchase Materials from Qualified Supplier. Using approved suppliers with approved quality control systems is the cost-effective way of preventing improper materials from entering the plant. Approved suppliers should provide the best overall package for the

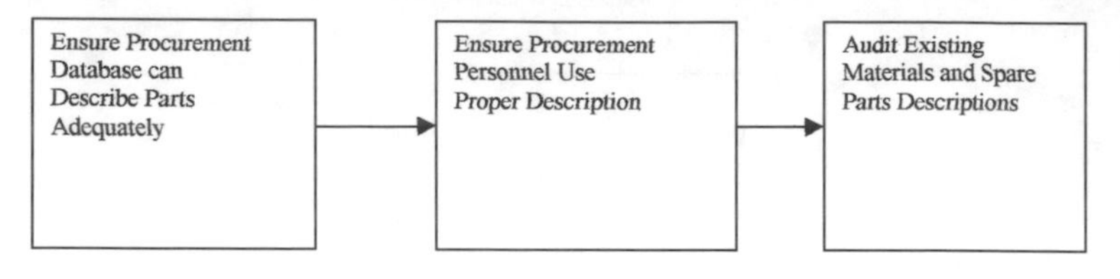

Figure 1-3. The procurement database must provide properly identified materials.

expenditure, warranties, quality programs, cost discounts, and technical assistance. See Figure 1-4.

Approved suppliers must be audited from time to time to verify they are maintaining their systems in preventing materials mix-ups. As a minimum requirement, approved suppliers should be ISO 9000 certified. Non-approved suppliers must also be audited, initially to assess the strengths and weaknesses of their quality program. Those suppliers who fail must improve their weak areas, or be dropped. See Figure 1-5.

Perform Basic Receiving Inspection on Incoming Materials. A basic receiving inspection verifies the quality of incoming materials and spare parts. Receiving inspections are best done at a supplier's warehouse before the materials are approved for shipment to the site. They are developed from applicable regional and national standards such as ASTM. Suppliers' quality control organizations should create appropriate programs and take responsibility for training their inspectors. Alternatively, the programs

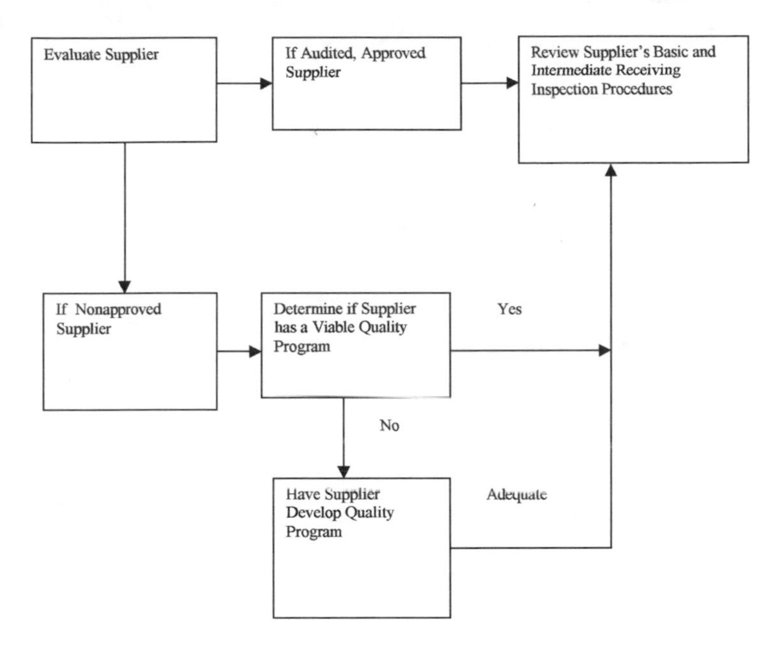

Figure 1-4. Materials suppliers must demonstrate their quality programs are acceptable.

may be set up at service centers. When neither situation exists, the plant receiving organization must absorb the cost of creating a program and training its own receiving inspectors. It is generally a less efficient approach than transferring inspection responsibility to the supplier. See Chapter 3 under Basic Receiving Inspections. See Figure 1-6.

Attribute	✓
Does supplier have a strong, documented quality assurance policy	
Is there a satisfactory acceptance process for sub-suppliers, such as a method of auditing their quality systems	
Do inspection systems follow applicable codes/standards or generally accepted principles	
Are internal order processing systems capable of conforming with purchasers' internal designation and procurement systems	
Do ordered components meet national or other standards, such as ASTM	
Are welders and welding procedures qualified and certified to meet ASME Code section 1X	

Figure 1-5. Materials suppliers must be audited for capability to meet purchasers' requirements.

> Basic receiving inspections check for adherence to specifications only. They are not intended to look for quality levels beyond which the materials are normally built. For example, "no oxidizable grease" in a chlorine or oxygen service valve would add an additional requirement over and above basic, making it a supplementary receiving inspection.

Perform Field Audit Before Use. The mechanic or craftsperson must verify that the part is indeed correct. Despite a seamless quality chain through to the receiving warehouse, the final delivery system could have switched parts. Thus, the mechanic or craftsperson is the "inspector of

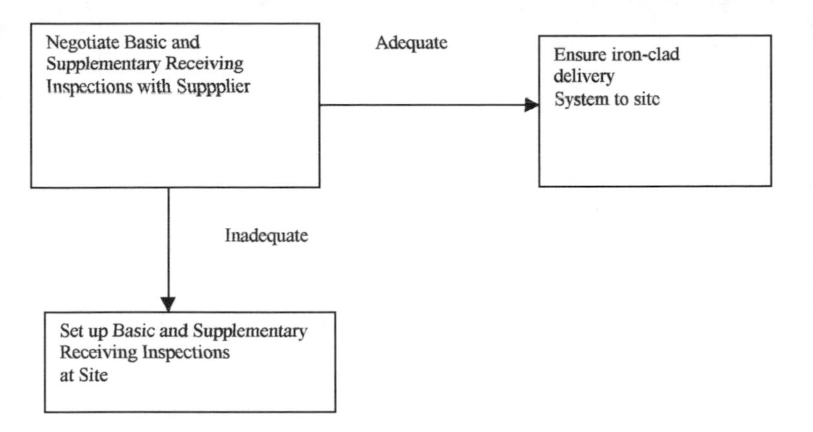

Figure 1-6. Basic and supplementary receiving inspections are required to check for adherence to specifications.

last chance." Discrepancies must be reported formally through the system for resolution. See Figure 1-7.

MANAGEMENT'S ROLE

Excellence in materials management requires continuous effort at all levels of the organization. Management's duty is to set up and sustain systems that avoid materials mix-ups through adherence to the following:

- Ensure procedures are in place to change specifications in accordance with field modifications

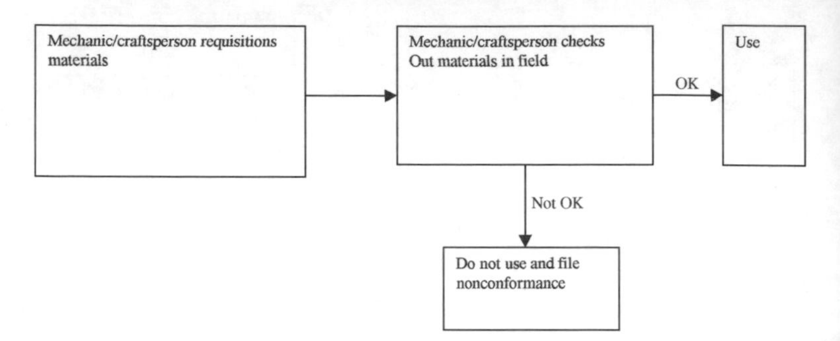

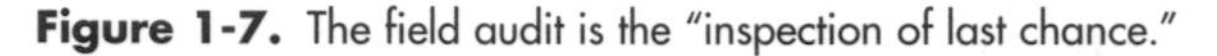

Figure 1-7. The field audit is the "inspection of last chance."

- Allow sufficient time for performance of process safety critical activities
- Support, communicate and recognize good practices, including the use of computerized maintenance management databases
- Schedule training and refresher training to maintain a properly skilled workforce
- Ensure a feedback system is in place for reporting discrepancies by all functions in the critical materials supply chain.
- Support, communicate and recognize standard operating procedures and operational discipline in the running of manufacturing plants

2
Specifying and Ordering Materials

Summary: This chapter describes how to use materials standards in a procurement specification, the types and acceptability of product certifications, the roles of responsible parties in the supply chain, and how to develop accurate computerized descriptions. Also discussed is the issue of replica parts.

Materials standards and codes are the common language between specifiers, buyers and producers. Materials standards and codes help ensure that materials meet the properties to which they are designed. Purchase orders must reference mechanical, physical and chemical property limits as defined by accepted standards.

Quality requirements are at the root of materials standards and codes. They are used mutually by manufacturers, suppliers and buyers for ordering and fabricating materials. Specifying and ordering materials are simplified by using materials standards and codes so that required performance characteristics may be achieved in service. See Figure 2-1.

INDUSTRY STANDARDS

Industry-wide standards include ASTM, AWS, ASME and SAE.

✓	Information Type
	Specification number (such as ASTM number or company standard number)
	Designation within the specification (such as ASTM grade number within specification or code number within company standard)
	Quality requirements (e.g., mechanical test results, chemical composition, special tests beyond those required in the specification and requiring the support of formal documentation)
	Dimensional requirements
	Tolerances
	Other requirements (such as surface finish, packing)
	Quantity
	Suggested supplier
	Reference drawing number (if applicable)

Figure 2-1. Information from the specification is the key to effective procurement of materials.

> A material is acceptable only if it meets the specified requirements outlined in the applicable materials standards.

ASTM. American Society for Testing and Materials produces the largest single source of standards, consisting of nearly 70 volumes divided by subject matter into 16 sections. The most commonly used ASTM standards in the process industry are available in a compendium, *ASTM Standards for Maintenance Repair and Operations Materials in the Chemical Process Industry.* ASTM uses letter-number designations to describe its standards. See Figure 2-2.

For example, A193 is a specification for bolting materials, but calling out A193 alone is not enough. Within each ASTM specification there are often several types of materials, known as embedded materials. For example, ASTM A193 for alloy steel and stainless steel bolting materials includes embedded materials such as:

- Grade B7 (high-strength, low-alloy steel)
- Grade B8 class 1 (304 stainless steel in the soft annealed condition)
- Grade B8 class 2 (304 stainless steel strengthened by cold work)

ASTM A193 also contains requirements for chemical composition, mechanical properties, steel making practice, manufacturing practice, and heat treatment practice. In addition, ASTM A193 details optional properties that may be

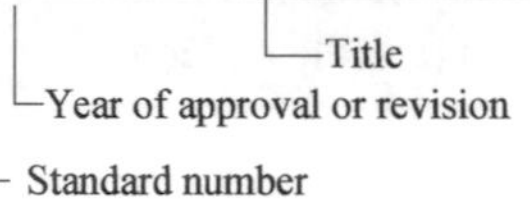

A = Ferrous metals

B = Nonferrous metals

C = Cementitious, ceramic, concrete, or masonry materials

D = Miscellaneous materials

E = Miscellaneous subjects

F = Materials for specific application

G = Corrosion, deterioration, degradation of materials

E = Emergency standards

P = Proposals

Figure 2-2. ASTM Standards provide a well ordered designation system.

required, such as impact strength testing. These are listed as supplementary requirements that the end user may select on an as needed basis, at extra cost. ASTM standards for products cover many industrial applications with varying quality requirements. Engineers select the standard with the applicable quality level for the product they require. For example, ASTM A53 is generic piping. A106 on the other hand is seamless (contains no weld) piping made to fine grain steelmaking practices (making it tougher). It is therefore less prone to leakage, making it more appropriate to be specified for critical service. Analysis of ASTM A53 and A106 reveals the key differences between the two standards. See Figure 2-3.

	ASTM A53	ASTM A106
Type of Standard	Specification	Specification
Title	Pipe, Steel, Black & Hot Dipped, Zinc Coated, Welded & Seamless	Seamless Carbon Steel Pipe for High Temperature Service
Commonly used type/grade within specification	Type E, Grade B	Grade B
Description	Electric resistance welded, slightly higher carbon (stronger) than Grade A. Not made to fine grain steelmaking practice	Seamless, good balance between strength & weldability. Made to fine grain steelmaking practice for more reliable properties
End use	Ordinary or general purpose	Critical service

Figure 2-3. Comparison of ASTM Standards allows the user to select the most appropriate standard for the application.

Indicate not only the ASTM specification number but also the embedded material indicated by the type, grade, or class designation when applicable. Never make arbitrary changes to ASTM specifications, including the embedded material designation without prior approval.

AWS Standards. American Welding Society produces standards on various aspects of welding and has developed a series of 31 standards on welding consumables, solders and brazes. They contain specifications on composition of the filler metal, composition of the completed weld deposit (when applicable), minimum strength of the completed weld, dimensional requirements of a weld, impact properties (when applicable) and radiographic standards. for example, specification A5.1 Carbon Steel Covered Electrodes includes the following types:

- E6010, a low strength electrode, which is deeply penetrating and with easily removed slag. It is sometimes used for the root pass in welding carbon steel piping.
- E7018, a higher strength electrode, which may be used in all positions and is low in hydrogen. It is commonly used for welding out carbon steel

See Figure 2-4. AWS publishes *Filler Metal Comparison Charts,* a book that compares trade names of welding products with standard designations.

Specify welding consumables, brazes and solders per AWS designation.

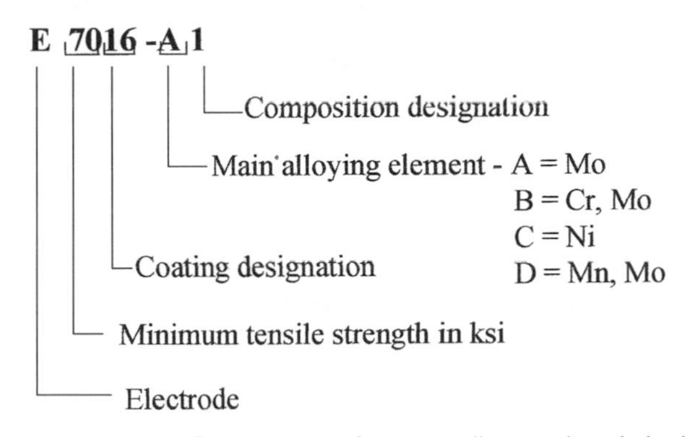

Figure 2-4. Identification system for AWS Filler Metals includes key information for the end user.

ASME Standards. American Society of Mechanical Engineers administers the Boiler and Pressure Vessel Code, which contains guidelines for the design and fabrication of boilers, unfired pressure vessels and other equipment. ASME does not develop its own material specifications but approves certain ASTM and AWS specifications. Minor changes are sometimes made prior to adoption of these specifications and they are given the prefix letter S to indicate ASME approval. For example, ASTM A516 becomes SA-516. Welding filler metals use the prefix letters SF. For example, AWS A5.1 covers carbon steel welding filler metals, whereas SF5.1 covers ASME code-approved carbon steel filler metals. Such materials must be procured for the fabrication or repair of code-stamped vessels. ASME also administers the design and fabrication codes for chemical plant and power piping, ASME B31.3 and ASME B31.1 respectively. These codes require use of materials made to ASTM standards.

SAE Standards. Society of Automotive Engineers standards cover specifications for highway and off-road vehicles Some SAE specifications may be used in chemical plant and refinery applications, e.g., bearing materials. SAE also administers the Aerospace Materials Specifications (AMS), which include procurement requirements for extreme quality aerospace materials. AMS products may be applicable for land based, highly stressed critical components such as reciprocating compressor shafts.

> Know when you should be ordering and using ASME code approved materials (e.g., for pressure vessel repair vs. ordering and using ASTM materials; for piping repairs).

METAL IDENTIFICATION SYSTEMS

The first orderly methods of identifying materials were developed by metal manufacturers trade associations. National standards organizations have also created materials identification systems. In the United States, the Unified Number System merges all systems into one method of identifying commercially available metals and alloys.

AISI. American Iron and Steel Institute produced designations for steels:

- For carbon and low alloy steel, a four digit system. The carbon content is usually indicated in the second two numbers and the alloy family in the first two numbers. For example, 1030 is a medium carbon steel containing nominally .3% carbon; 4140 is a low alloy steel containing nominally 1% chromium and .25% molybdenum with .4% carbon. See Figure 2-5.
- For stainless steels, a three digit system sometimes followed by a letter. For example, 304L stainless steel is an extra low carbon version of 304 stainless steel. Steels in the 300 designation are known as austenitic stainless steels.
- For tool steels, a letter followed by one or two numbers. For example, D2 belongs to a family of high carbon, high chromium tool steels known as cold work tool steels. All members of this family have the prefix D.

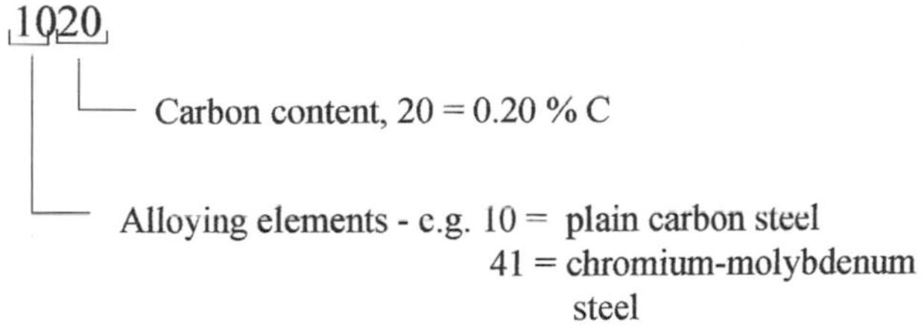

1020

Carbon content, 20 = 0.20 % C

Alloying elements - e.g. 10 = plain carbon steel
41 = chromium-molybdenum steel

Figure 2-5. AISI Designation System for carbon and low alloy steels consists of four numbers.

CDA. Copper Development Association produced designations for copper alloys using a three-digit system. For example, copper alloy 836 is a cast leaded brass containing specified amounts of zinc, lead and tin.

AA. Aluminum Association produced a four-digit system for wrought aluminum alloys and a three-digit system for cast alloys. Examples are alloys 6010 (wrought) and 356 (cast). A temper designation is also required. It follows the alloy designation and indicates the heat-treated or mechanically worked condition of the alloy such as 6010-T6. The T6 suffix indicates the alloy is in the precipitation hardened condition for optimum strength.

ACI. Alloy Castings Institute produced a system for corrosion resistant and heat resistant castings. The letter C indicates the corrosion series and the letter H indicates the heat series. For example, CF-8 is a corrosion resistant stainless steel and HK-40 is a heat resistant stainless steel.

API. American Petroleum Institute writes specifications for carbon and alloy steel piping used in the oil and gas industry, which comprises line pipe, drill pipe, casing and tubing. These are covered in API specifications 5CT (casing and tubing), 5D (drill pipe) and 5L (line pipe). Contained in these specifications are various embedded materials grades. See Figure 2-6.

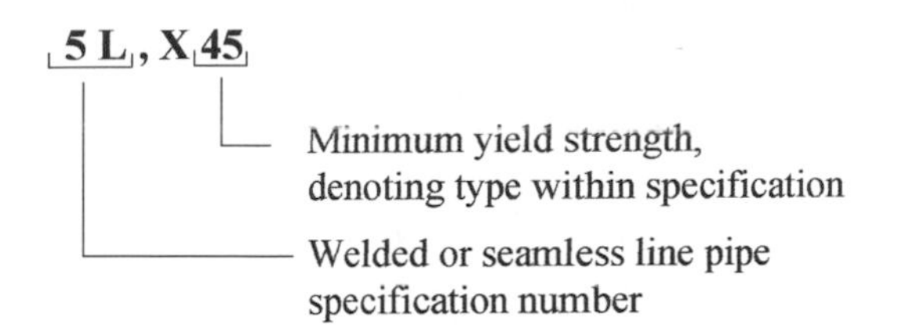

Figure 2-6. API specifications cover casing and tubing, drill pipe, and line pipe.

CSA. Canadian Standards Association designations for carbon and alloy steel pipe are contained in CSA specification CA3-Z245.1.

DIN. Deutsch Institut fur Normung develops standards for materials in Germany. Two systems are used to describe metals by their chemical composition. The first consists of the letters DIN followed by an alphanumeric or numeric code. The second, known as the Werkstoff number, uses numbers only with a decimal point after the first digit. See Figure 2-7.

UNS. Unified Numbering System satisfies the need for a common designation system for all alloys. The UNS number uniquely identifies the chemical composition of alloys that have been fixed by other organizations, such as one of the metal manufacturers associations. If the alloy is proprietary or produced by a limited number of suppliers, the chemical composition is established by the supplier. The

MATERIAL NUMBER (Werkstoff Number)

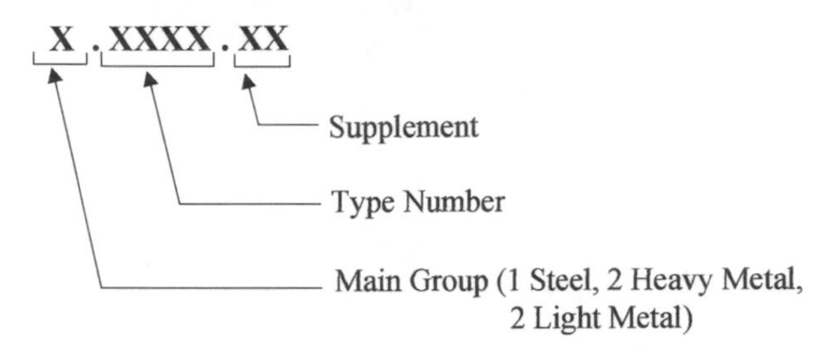

e.g.: 316 stainless steel is W. No. 1.4401
316Ti stainless steel is W. No. 1.4571
See DIN 17007 for details

MATERIAL SYMBOL

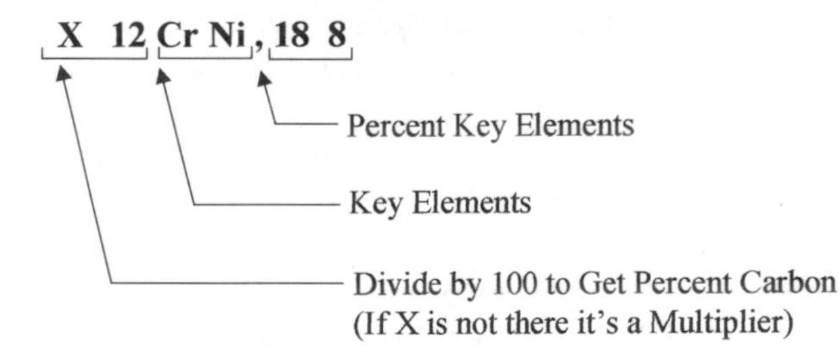

e.g., X 6 Cr Ni Mo Ti 17 12 2 = 1.4571 = 316 Ti

See DIN 17006 for details

Figure 2-7. German designation for metals employ two systems.

UNS consists of a letter followed by five numbers. The letter identifies the alloy family and, where possible, the five numbers are related to the pre-UNS designation of the alloy. For example, copper alloy C61400 was formerly CDA 614 and S30400 was formerly AISI 304 stainless steel. See Figure 2-8.

UNS publishes a book, *Metals and Alloys in the UNS,* that cross-references all alloys having a UNS designation with ASTM, ASME and other specification systems in which each specific alloy is referenced. See Figure 2-9.

> Specifying UNS designation alone is not sufficient. One must also reference the product specification, such as ASTM number.

MATERIALS TEST REPORT (MTR), CERTIFICATE OF COMPLIANCE (COC) AND PRODUCT ANALYSIS

MTR, COC and Product Analysis are different types of paperwork that suppliers will provide to customers to qualify their materials. See Figure 2-10.

MTR is a certified statement issued by the primary manufacturer of a metal product (the mill) indicating the chemical analysis and mechanical properties of the stock, such as plate, sheet, strip or bar. MTR's are sometimes known as Certificates of Analysis (COA). MTR's are not formally required for all types of ASME code approved materials that

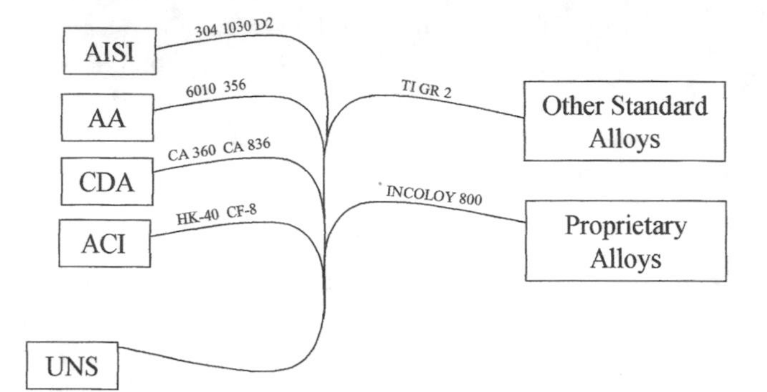

AXXXXX	Aluminum and Aluminum Alloys
CXXXXX	Copper and Copper Alloys
EXXXXX	Rare Earth and Similar Metals and Alloys
FXXXXX	Cast irons
GXXXXX	AISI and SAE Carbon and Alloy Sheets
HXXXXX	AISI and SAE H-Steels
JXXXXX	Cast Steels (Except Tool Steels)
KXXXXX	Miscellaneous Steels and Ferrous Alloys
LXXXXX	Low Melting Metals and Alloys
MXXXXX	Miscellaneous Nonferrous Metals and Alloys
NXXXXX	Nickel and Nickel Alloys
PXXXXX	Precious Metals and Alloys
RXXXXX	Reactive and Refractory Metals and Alloys
SXXXXX	Heat and Corrosion Resistant Steels (Including Stainless), Valve Steels, and Iron-Base "Superalloys"
TXXXXX	Tool Steels, Wrought and Cast
WXXXXX	Welding Filler Metals
ZXXXXX	Zinc and Zinc Alloys

Figure 2-8. Unified Numbering System satisfies the need in the United States for a common designation system for all alloys.

DIFFERENT 304 SST SPECIFICATIONS

Chemical Composition	Specifications
C 0.8 max, Cr 18.00-20.00, Mn 2.00 max, Ni 8.00-10.50, P 0.045 max, S 0.030 max, Si 1.00 max	AISI 304 UNS S30400 AMS 5501;5513;5560;5563;5564; 5565;5566;5567;5639;5697;7228;7245 ASME SA182 (304); SA194; (B8); SA213 (304); SA240 (304); SA249 (304); SA312 (304); SA320 (B8); SA358 (304); SA376 (304); SA403 (304); SA409 (304); SA430 (304); SA479 (304); SA688 (304) ASTM A167 (304); A182 (304); A193 (B8); A194 (B8); A213 (304); A240 (304); A249 (304); A269 (304);A270 (304); A271 (304); A276 (304); A312 (304); A313 (304); A314 (304); A320 (B8); A358 (304); A368 (304); A376 (304); A409 (304); A430 (304); A473 (304); A478 (304); A479 (304); A492 (304); A493 (304); A511 (304); A554 (304); A580 (304); A632 (304); A666 (304); A688 (304) SAE J405 (30304)

Figure 2-9. UNS designation describes the chemical composition of alloys that may be specified in several systems.

are to be used for code work, such as a repair, but many types of companies insist on it.

MTR's allow the end user or receiving organization to cross-check that the product meets chemical composition and mechanical property requirements of the specification,

COMPONENT	MTR	COC*
Fasteners	Difficult to relate to product**	May be meaningful
Gaskets	Not applicable	May be meaningful
Valves	Only covers body castings	May be meaningful
O-Rings	Not applicable	May be meaningful
Tube and Pipe	Is meaningful	May be meaningful
Pipe Fittings and Flanges	Is meaningful	May be meaningful
Plate, Sheet and Strip	Is meaningful	May be meaningful
Welding Consumables	Is meaningful	May be meaningful
Valve Repair Kit	Not likely to be meaningful	May be meaningful
Barstock, Threaded Stock	Is meaningful	May be meaningful
Forgings	Is meaningful	May be meaningful

*Depends on knowledge of supplier's quality system.
**Depends if representative lot is produced.

Figure 2-10. Mill test reports and certificates of compliance have distinct areas of applicability.

for example by ASTM and other organizations. The information in an MTR is arranged for ease of interpretation. See Figure 2-11.

> MTR's are normally free, but in some cases may incur a cost. For example, the rod from which a batch of fasteners is made may reference an applicable MTR. If the rod having different MTR's is mixed to create one large lot of fasteners, as often happens, it is not possible to assign a specific MTR to an individual fastener or group of fasteners. If the end user requires an MTR for incoming fasteners, the lot of material corresponding to a specific MTR must be separated and kept separate during the entire manufacturing process. This adds to cost, which is passed on to the user.

COC is a statement by a manufacturer or supplier that a material meets specifications, but it does not contain supporting, quantitative documentation.

COC's can be issued for all materials. In the case of fasteners, COC's would state that the original rod met the referenced specification and the manufacturing process complied with the purchase order (e.g., rolled threads vs. machined). A COC contains no test reports (e.g., if a gasket was made from Teflon® the COC should state it was made of Teflon® and that it is being supplied). No tests were made to determine that it was Teflon®; it only states that the records indicate the supplier is confident that no substitutions have been made.

ABC Company
Houston, TX
Reference Number:
Customer name: Lynco Flange & Fittings, Inc.

Date: 07/27/00

Item Data

Item Description	Heat Code	Heat Number
1" 150 Lap Joint SA105	4H7	38917

Chemical Properties

Heat No.	C	Si	Mn	P	S	Cr	Al	Cu	Ni	Mo	V	Cb				CE
4H7	0.180	0.220	0.860	0.009	0.014	0.080	0.034	0.240	0.120	0.020	0.003	0.004	0.000	0.000	0.000	0.368

Physical Properties

Yield Strength	Tensille Strength	Elong.	Red. Area	Hardness BHN	Charpy Test	Foot Pounds	Lat. Expan.	Shear Frac.	Test Temp
46,110	74,675	29.00	68.30	150-160					

Notes

The information contained herein has been electronically transmitted to our customers.

We certify our flanges are capable of passing hydrostatic test compatible with their rating.
We certify that all test results and process information contained herein are correct and true as contained in the records of the company.

Figure 2-11. Information in a mill test report is easy to interpret.

Product Analysis is a chemical report that a particular product form, such as tubing or pipe fitting, is made from a specific heat of metal. Product analyses ensure that substitutions have not been made during processing the metal. Product analyses are called out as supplemental requirements in ASTM specifications and cost extra, unlike the MTR or COC.

> Despite MTR's and COC's, non-conformances can still occur. However, they put the supplier on notice that the receiving site has value for controlling incoming materials. Before requesting MTR's and COC's, answer the following questions:
>
> - *Does the MTR qualify the actual component? For example, an MTR for a valve is available only for the body casting and the wetted parts such as the bonnet. The remaining parts, such as the stem, seat etc, would have to be documented on a COC unless specifically requested.*
> - *Is the COC from a reputable supplier? If not, it is better to optimize your product supply chain so what is specified is actually obtained, irrespective of the information provided in a COC.*

ORDERING MATERIALS

Ordering requires adherence to the standard. Different types of commodities require different amounts of information for adequate specification. The more complex the commodity, the greater the amount of information required. A valve, for example, requires significantly more information than a fastener. Chapter 3 indicates minimum types of information required to specify various commodities. See Figure 2-12.

> Wherever possible, use industry-wide specifications, such as ASTM, because they result in the lowest cost product meeting the desired level of quality for the service. Internal specifications may be required for more complex commodities such as valves, or for gaskets where no industry-wide specifications exist.

COMPANY STANDARDS	INDUSTRY STANDARDS
Gaskets	Fasteners
Valves	Pipe Fittings
Packing	Flanges
O-rings	Plate
Piping Systems	Sheet & Strip
	Tube & Pipe
	Welding Consumables

Figure 2-12. Companies must develop internal designations for commodities such as valves where no industry designations exist.

REPLICA PARTS

Replica parts are substitutions for spares provided by the original equipment manufacturer (OEM) and usually obtained at lower cost. They are obtained from suppliers who specialize in replica parts or they may be manufactured by the end user when an accurate drawing is available.

Replica parts programs must be conducted with care. It is not adequate to merely substitute an equivalent alloy based on chemical analysis. The stock for the replica part must be purchased in the form (e.g., forged vs. cast) and condition (e.g. quenched and tempered vs. normalized) that meets the required quality level in the component. For example, rolled plate cannot be substituted for a forging without evaluation of the design and operating conditions. When it comes to replica parts made of hard materials, plastics or rubber, the confidence level is significantly less than with metals. This is because even subtle changes in the formulation between supposedly equivalent materials may result in significantly different properties. See Figure 2-13.

Before embarking on a replica parts program, the design group must ensure that an improper substitution is not made. Furthermore, replica parts usage may void the warranty. It is important that a replica parts program makes both economic and technical sense. Renegotiation of the cost of OEM parts with the manufacturer is always an option.

Item	Example
Part Name	Auger: a. shaft b. flights
Criticality	Not highly loaded or high speed, or requiring special considerations
Effect on Warranty or Liability	Does not void warranty
Ability to Meet Dimensions and Shape	Can be achieved by machining
Acceptable Processing Method	Machining of wrought stock
Current Material	a. DIN C15 b. DIN U St 37-1
Equivalent material	a. 1015 carbon steel b. 1013 carbon steel
Acceptable Replacement Material	a. 1020 carbon steel rolled b. 1020 carbon steel rolled

Figure 2-13. Replicating parts of original equipment manufacturers requires careful assessment and use of knowledgeable resources.

DATABASE DISCIPLINE

A computerized materials management database is only as good as the information that can be retrieved from it. Success depends on everyone using the same procedure, key nouns and correct equipment piece numbers. Otherwise, bookkeeping and business decisions suffer. For example, in one case a plant was shut down for three hours because the data search person and the data entry person used different descriptions for a critical part.

There are no gray areas with computers. They demand a greater degree of discipline than manual data entry.

> When you sit down at the computer have all the correct information—item number, type, grade, class of materials, special requirements, etc. Complete all essential fields exactly as required. Even if you think you know a better way or shortcut, someone else might not. If you don't know a procedure, look it up, or get training. See Figure 2-14.

Computer Commandments

1. Be precise—computers can't interpret
2. Enter data consistently—no shortcuts
3. Know your own correct key nouns—no inventions!
4. Enter data yourself or guide your data entry person
5. Double check entry for accuracy—spelling, spacing, tax terminology
6. Track hours spent accurately—no fudging!
7. Relate bill of materials to major equipment item or sytem—a pump may have 200 parts
8. Research the part number or company code number thoroughly
9. Enter accurate piece number on work order
10. Enter complete field descriptions—include locations, avoid slang

---

Figure 2-14. Computerized systems require a greater degree of discipline than manual data entry.

Maintenance and repair costs (M&R) are a deductible corporate expense against taxable income. Using appropriate tax terminology on all M&R work orders will save your company money. The costs associated with Capital Expenditures (capital) and Service to Operations are not deductible. Key words used in computerized materials management databases should help clarify these categories and should be used for effective cost savings. See Figure 2-15.

SPECIFYING/ORDERING SUPPLY CHAIN—ROLES AND RESPONSIBILITIES

The materials procurement supply chain contains many handoffs, making it is difficult to control the quality and accuracy of items. See Figure 1-2.

Job function responsibilities are as follows:

Specification (engineers):

- Call out materials and develop cost-effective, current specifications
- Use failure frequency information/relevant data to determine when materials failure analysis is required
- Keep abreast of latest changes in regulations and codes
- Know process changes that might affect types of materials used in the operating area
- Work with other functional groups to make sure specifications are updated as field changes are made
- Know expert resources and how to contact them
- Organize inspection programs for commodities as needed

Maintenance & Repair	Capital Expenditures	Service to Operations
Adjust	Alter	Assist in process
Calibrate	Assemble	Clean process equipment
Change	Construct	Handle process material
Check	Enlarge	Paint
Dismantle	Extend	
Inspect	Fabricate	
Insulate	Improve	
Lubricate	Increase capacity	
Maintain	Install	
Patch	Inspect & replace	
Refinish	Make	
Regasket	Make alterations	
Remove	Modify	
Renew	Pull & overhaul	
Repack	Purchase & install	
Repair	Purchase & replace	
Replace*	Rebuild	
Rotate	Reroute	
Sandblast	Revamp	
Service equipment	Revise	
Test		
Tighten		

*Replacements are charged as repairs only if the replacement represents a small percentage of the fixed asset or if the replacement does not adapt the item to a different use, does not appreciably prolong its life or materially increase its value. For example, replace pumps, seals, gaskets, etc. is a cost associated with maintenance and repair (M&R).

Figure 2-15. Use of appropriate maintenance and repair cost tax terminology on all maintenance will save money since they are deductible expenses against income.

Procurement (materials clerks, planners, stores, local buyers):

- Acquire, order, or oversee requisition of materials and ensure the bill of materials (BOM) is complete and accurate
- Ensure procurement practices meet company guidelines
- Keep pricing in line
- Distribute or oversee distribution of correct parts to pertinent people (mechanics/operators)
- Ensure critical service items are inspected and verified and meet specifications
- Properly store and label all materials using bar coding if possible and additional bin labeling in case of computer outage
- Oversee accuracy of contractors

Supply (vendors, manufacturers' representatives, component rebuilders):

- Ensure that materials and components meet applicable codes, specifications and certification
- Match quality level with criticality of service
- Ensure quality with third party suppliers

3

Receiving and Verifying Materials

Summary: This chapter indicates the information to check and the inspections to make when receiving parts to ensure they meet purchase specification requirements. The types of items covered are fasteners; pipe fittings and flanges; pipe and tubing; plate, sheet and strip; valves; gaskets; o-rings; and welding consumables. Responsibilities of the various roles in the receiving organization are described.

Incoming materials must be checked for conformance with the specification and purchase order, and then labeled accordingly. Materials received and stored must be compared with the purchase order and accompanying paperwork. Mechanics and crafts people must conduct a field audit of the materials before putting them into service.

RECEIVING INSPECTION STEPS

The following types of receiving inspection steps help minimize materials mix-ups:

Sampling. Receiving inspections usually involve a sampling procedure (e.g., for small items such as gaskets, 5% of the lot of the product is sampled). If a piece within the sample is nonconforming, the lot is either rejected or 100% inspection is performed. For larger or more complex items, such as valves or flanges, 100% inspection might be performed. All components for process safety critical applications may require blanket 100% inspection.

> Learn how to use the sampling system at your receiving location.

Certification. The various types of paperwork used to check products include materials test report (MTR), certificate of compliance (COC), product analysis and purchase order. See Chapter 2 under MTR, COC and Product Analysis. The product description should match the purchase order.

> Learn to read standards such as ASTM, interpret mill test reports and product analyses, and understand when certificates of compliance are applicable. Learn the differences between embedded materials within standards and supplementary requirements in standards. See Chapter 2 for further information.

Chemistry. Chemical analysis methods range all the way from simple to sophisticated. See Chapter 6 under Field Identification. The exact chemical composition is generally not required. The identification method should distinguish the correct material from materials that may have been substituted. For example, a moly (molybdenum) spot test is a chemical color change technique that is sometimes used to distinguish 304 (no moly) from 316 (small percentage of moly) stainless steel.

Physical properties. The most common physical property test is a magnet. It may be used when it can distinguish correct material from materials that may have been substituted. For example, a magnet will distinguish 300 series stainless steel from 400 series when it is known that a valve seat is either 304 or 410 stainless steel.

Learn how to use basic analytical techniques recommended by your inspection group. See Chapter 6.

Mechanical properties. Hardness is the most common mechanical test. See Chapter 5, Mechanical Testing. Hardness testing, for instance, may be used to indicate whether fasteners in specific critical services are greater or less than the value required to avoid a premature failure. Mechanical tests other than hardness are destructive. If a destructive test is required to ensure compliance it is performed on a sample piece from the lot. Mechanical tests are time consuming and only used when required by the specification.

> Obtain guidance on the applicability or need for mechanical tests from your inspection group. See Chapter 6.

Appearance. Verify parts by understanding their marking system and develop a trained eye. See Chapter 6, Identification Markings. The mechanic or planner learns how a material should "look." If a nut, for example, is made by powder metallurgy (P/M) it may be porous and lack the corrosion resistance of an equivalent wrought product. Close examination of a P/M nut shows a surface full of tiny pores, whereas a wrought nut will have an unbroken surface. Basic problems such as imperfections on flange faces must be identified and the cause of rejection documented.

> Learn how critical components are fabricated and identified to help you understand appearance factors. See Chapters 5 and 7.

Measurement. Simple tools such as calipers, micro-meter or the trained eye are used to ensure suitability for service. With certain types of commodities it is easy to set up a "go/no-go" gauge to compare measurements with the specification. For example, checking gaskets on a pegboard containing standard sizes or o-rings on a conical gage for internal diameter may be acceptable methods in your receiving organization.

> Learn the key dimensions of the critical commodities that your site or area purchases.

Documenting Conformance vs. Nonconformance. For nonconforming product the purchase or sales order is completed, initialed and dated. Nonconforming product must be recorded on the applicable documentation, such as a Materials Exception Report (MER), or an Overage, Shortage and Damage Report (OSDR). Nonconforming items are moved to a separate nonconforming product area. A receipt inspection verification report is a related form that is used identify the method of inspection and provide disposition and follow-up for closure. See Figure 3-1.

> Document not only nonconformance but also causes of nonconformance so that problems may be passed back to purchasing in order to eliminate them from your materials supply chain.

Labeling and Tagging. Conforming product must be properly labeled and tagged. Tags must accompany critical materials throughout storage and assembly. Should tags be lost, reinspection and new tags are required. Materials and storage bins require identification labels to enable accurate storage and disbursement. There must be adequate protection from the environment, especially if prolonged shelf life is indicated. See Figure 3-2.

BOLTS, STUD BOLTS, STUDS AND CAP SCREWS

Applicable Standards. Examples of fastener standards for process plants include the following:

Materials Exception/Verification Report

Purchase Order Number ______ Process Safety Management Critical? Yes No

Project/Work Order Number ____ Unit/Area __________________

Item Identification Number ____ or Description ________________

Manufacturer or Supplier Identification/Name _____________________

Inspection Aspect	Accept	Reject	N/A	Date Closed
1. Item Identification Number	______	______	______	______
2. Per Purchase Order Specification	______	______	______	______
3. Quantity	______	______	______	______
4. Damage	______	______	______	______
5. Supporting Technical Literature	______	______	______	______
6. Protection	______	______	______	______
7. PMI of Applicable Components	______	______	______	______
8. Materials Test Reports (MTR's)	______	______	______	______
9. Certificates of Compliance (COC's)	______	______	______	______
10. Manufacturer's ID Number	______	______	______	______
11. Item in Correct Storage Area	______	______	______	______

Figure 3-1. Quality problems must be documented so that they may be passed back through the supply chain.

- ASTM A193, "Alloy Steel Bolting Materials for High Temperature Service"
- ASTM A307, "Carbon Steel Bolts and Studs, 60,000 Tensile Strength
- ASTM A320, " Alloy Steel Bolting Materials for Low Temperature Service"

✓	Checklist
	Item number
	Critical indicator flag
	Location number
	Description (short)
	Unit of issue
	Order point
	Manufacturer
	Manufacturer's number
	Material of construction
	Drawing number, if appropriate

Figure 3-2. Labels or tags must provide adequate documentation of the material.

- ASTM A325, High Strength Bolts for Structural Steel Joints"

Minimum Information To Specify. The following information should be entered in the procurement database:

- Applicable ASTM standard number
- Embedded material (grade/class/type) within standard
- Type (e.g., stud)

- Thread standard (when applicable)
- Length
- Diameter

Basic Receiving Inspections. The following inspection steps are required:

- Sampling. Per guidance of the engineering organization.
- Certification. Check box to verify the part number matches what was ordered. If not, ensure it is an acceptable substitute. It is not normal practice to receive MTR's with fasteners.
- Appearance. Check head markings for fastener type and manufacturer. See Figure 3-3. Fasteners should be closely checked for manufacturing or finishing imperfections. Cracks and thread damage are unacceptable. Laps are acceptable if they pass the minumum requirements of ASTM F788, "Specification for Surface Discontinuities of Bolts, Screws and Studs." See Figure 3-4.
- Measurement. Studs are measured to first full thread. Bolts are measured tip to under head. Use go/no-go gauge to check diameter. See Figure 3-5.

Supplementary Receiving Inspections. The following additional inspections may be necessary to meet

(text continued on page 50)

Figure 3-3. Fastener head markings indicate fastener type and additionally bear imprint of the manufacturer's logo.

Identification grade mark	Specification	Fastener description and service	Material
No grade mark	SAE J429 Grade 1	Bolts, screws, studs	Low or medium carbon steel
	SAE J429 Grade 2		
No grade mark	SAE J429 Grade 4	Studs	Medium carbon cold drawn steel
 307A 307B	ASTM A 307 Grades A & B	Bolts screws, studs	Low carbon steel
	SAE J429 Grade 5	Bolts, screws, studs	Medium carbon steel, quenched and tempered
	ASTM A 449		
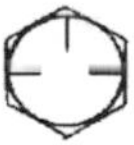	SAE J429 Grade 5.1	Sems[a], screws, studs	Low or medium carbon steel, quenched and tempered
	SAE J429 Grade 5.2	Bolts, screws, studs	Low carbon martensitic steel, quenched and tempered
	SAE J429 Grade 7 See Note 4	Bolts, screws	Medium carbon alloy steel, quenched and tempered
	SAE J429 Grade 8	Bolts, screws, studs	Medium carbon alloy steel, quenched and tempered
No grade mark	SAE J429 Grade 8.1	Studs	Medium carbon alloy or SAE 1041 modified elevated temperature drawn steel
	ASTM A 574	Alloy socket head cap screws	

Figure 3-3.
Continued.

Identification grade mark	Specification	Fastener description and service	Material
4.6	SAE J1199 Class 4.6	Metric bolts, screws, studs	Low or medium carbon steel
4.8	SAE J1199 Class 4.8		
5.8	SAE J1199 Class 5.8		
8.8	SAE J1199 Class 8.8		Medium carbon steel, quenched and tempered; may be alloy
9.8	SAE J1199 Class 9.8		
10.9	SAE J1199 Class 10.9		Alloy steel, quenched and tempered
B5	ASTM A 193 Grade B5	Bolts, screws, studs for high-temperature service	AISI 501
B6	ASTM A 193 Grade B6		AISI 410
B7	ASTM A 193 Grade B7		AISI 4140, 4142, or 4145
<u>B7M</u>	ASTM A 193 Grade B7M See Note 1		AISI 4140H, 4142H, or 4145H
B16	ASTM A 193 Grade B16		CrMoVa alloy steel

Figure 3-3.
Continued.

Identification grade mark	Specification	Fastener description and service	Material
B8	ASTM A 193 Grade B8	Bolts, screws, studs for high-temperature service	AISI 304
B8C	ASTM A 193 Grade B8C		AISI 347
B8M	ASTM A 193 Grade B8M		AISI 316
B8T	ASTM A 193 Grade B8T		AISI 321
B8	ASTM A 193 Grade B8 Class 2 See Note 2		AISI 304 strain hardened
B8C	ASTM A 193 Grade B8C Class 2 See Note 2		AISI 347 strain hardened
B8T	ASTM A 193 Grade B8T Class 2 See Note 2		AISI 321 strain hardened
B8M	ASTM A 193 Grade B8M Class 2 See Note 2		AISI 316 strain hardened
B8S	ASTM A 193 Grade B8S		UNS 521800
B9G	ASTM A 193 Grade B8M2 Class 2B		AISI 316 strain hardened

Figure 3-3.
Continued.

Identification grade mark	Specification	Fastener description and service	Material
L7	ASTM A 320 Grade L7	Bolts, screws, studs for low-temperature service	AISI 4140, 4142, or 4145
L7M	ASTM A 320 Grade L7M See Note 3		
L7A	ASTM A 320 Grade L7A		AISI 4037
L7B	ASTM A 320 Grade L7B		AISI 4137
L7C	ASTM A 320 Grade L7C		AISI 8740
L43	ASTM A 320 Grade L43		AISI 4340
B8	ASTM A 320 Grade B8		AISI 304
B8C	ASTM A 320 Grade B8C		AISI 347
B8F	ASTM A 320 Grade B8F		AISI 303 or 303Se
B8M	ASTM A 320 Grade B8M		AISI 316
B8T	ASTM A 320 Grade B8T		AISI 321

Figure 3-3. Continued.

Identification grade mark	Specification	Fastener description and service	Material
B8	ASTM A 320 Grade B8 Class 2 See Note 8	Bolts, screws, studs for low-temperature service	AISI 304 strain hardened
B8C	ASTM A 320 Grade B8C Class 2 See Note 8		AISI 347 strain hardened
B8F	ASTM A 320 Grade B8F Class 2 See Note 8		AISI 303 or 303Se strain hardened
B8T	ASTM A 320 Grade B8T Class 2 See Note 8		AISI 321 strain hardened
B8M	ASTM A 320 Grade B8M Class 2 See Note 8		AISI 316 strain hardened
BC	ASTM A 354 Grade BC	Bolts, studs	Alloy steel, quenched and tempered
BD	ASTM A 354 Grade BD See Note 5		
A325	ASTM A 325 Type 1	High-strength structural bolts	Medium carbon steel, quenched and tempered
A325	ASTM A 325 Type 2		Low carbon martensitic steel, quenched and tempered

Identification grade mark	Specification	Fastener description and service	Material
A325	ASTM A 325 Type 3 See Note 3	High-strength structural bolts	Atmospheric corrosion resisting steel, quenched and tempered
A490	ASTM A 490	High-strength structural bolts	Alloy steel, quenched and tempered

Notes:

1. ASTM specifications:
   A 193—Alloy-Steel Bolting Materials for High-Temperature Service
   A 307—Low-Carbon Steel Externally and Internally Threaded Standard Fasteners
   A 320—Alloy-Steel Bolting Materials for Low-Temperature Service
   A 325—High-Strength Bolts for Structural Steel Joints, Including Suitable Nuts and Plain Hardened Washers
   A 354—Quenched and Tempered Alloy Steel Bolts and Studs with Suitable Nuts
   A 449—Quenched and Tempered Alloy Steel Bolts and Studs
   A 490—Quenched and Tempered Alloy Bolts for Structural Steel Joints
   A 574—Alloy Steel Socket-Head Cap Screws
   F 568—Carbon and Alloy Steel Externally Threaded Metric Fasteners
2. SAE specifications:
   J429—Mechanical and Material Requirements for Externally Threaded Fasteners
   J1199—Mechanical and Material Requirements for Metric Externally Threaded Steel Fasteners
3. For metric classes 4.6, 4.8, 5.8, 8.8, 9.8, and 10.9; ASTM A 568 and SAE J1199 are equivalent.
4. Grade 7 bolts and screws are roll threaded after heat treatment.
5. A 354 Grade BD in sizes 1/4 through 1-1/2 must be marked with 6 radial lines 60° apart and may be marked with symbol BD; other sizes marked with symbol BD.
6. Marking symbols for identifying the manufacturer are optional or required depending upon specification.

1. B7M and L7M require 100% hardness testing. Underline indicates this.
2. The underline indicates these fasteners are Class 2, strain hardened.
3. The underline indicates this fastener is Type 3, a special weathering alloy.

Figure 3-3. Continued.

(text continued from page 44)

specific requirements of the purchase order, e.g., for critical applications:

- Field Identification. Supplemental chemistry verification by x-ray fluoresence spectroscopy or chemical spot test (e.g., for moly in 316 stainless steel) may be required.
- Mechanical Properties. Hardness (e.g., for compliance with ASTM A193 grade B7M).

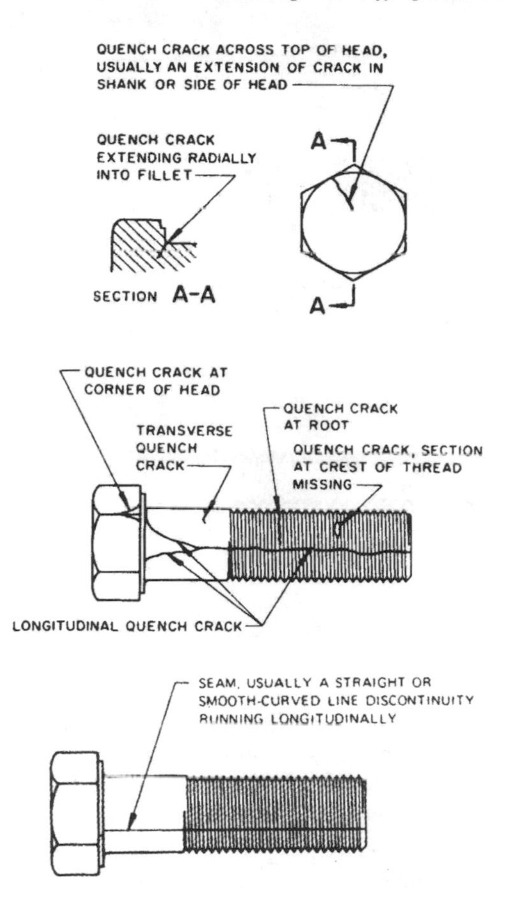

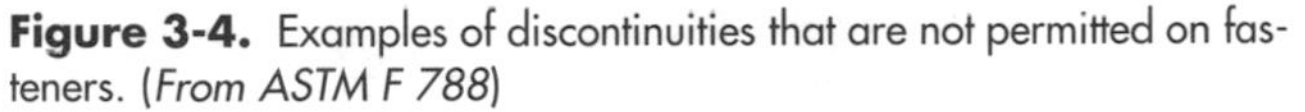

Figure 3-4. Examples of discontinuities that are not permitted on fasteners. (*From ASTM F 788*)

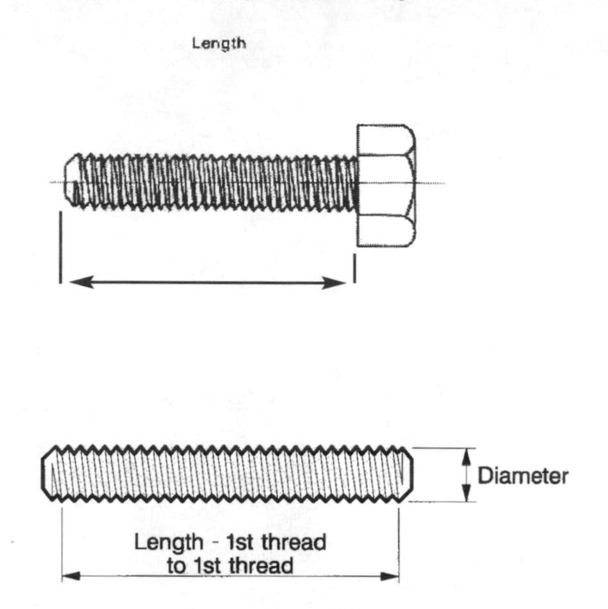

Figure 3-5. Bolts (upper) and studs (lower) are measured to different points.

- Considerations for Galvanized Fasteners. Zinc-coated fasteners for corrosion protection may be of two types: hot dip galvanized vs. electro-galvanized. Hot dip galvanized nuts are tapped oversize so as to compensate for the thicker coating on the bolt or stud compared with an electrogalvanized coating which is thinner. *Do not mix hot dip galvanized and electrogalvanized fasteners in bins.*

NUTS

Applicable Standards. Examples of nut standards in the process plants include the following:

- ASTM A194, "Carbon and Alloy Steel Nuts for Bolts for High Pressure and High Temperature Service."
- ASTM A563, "Carbon and Alloy Steel Nuts."
- ASTM A563, "Alloy Steel Bolting Materials for Low Temperature Service." This is the most generic specification for nuts, which neither requires marking nor is ASME code-aproved.

Minimum Information To Specify

- Applicable ASTM standard number
- Embedded material (grade/class/type) within standard
- Type (Nut)
- Thread standard (when applicable)

Basic Receiving Inspections. The following inspection steps are required:

- Sampling. Per guidance of the engineering organization.
- Certification. Check box to verify the part number matches what was ordered. It is not normal practice to receive MTR's with fastener materials.
- Appearance. Check markings for nut type and manufacturer. Check for imperfections—cracks and thread damage are unacceptable. Laps are acceptable if they

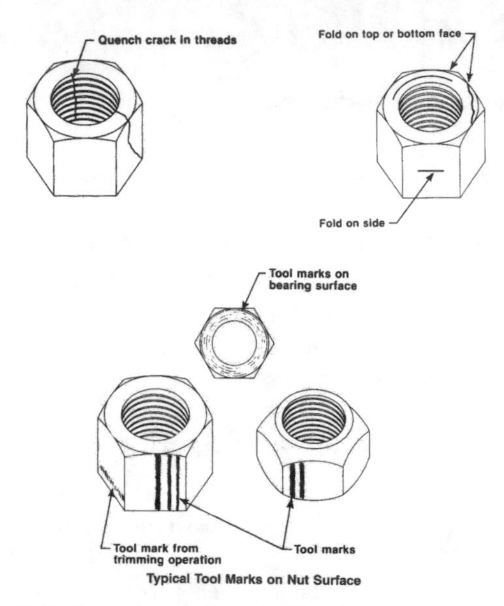

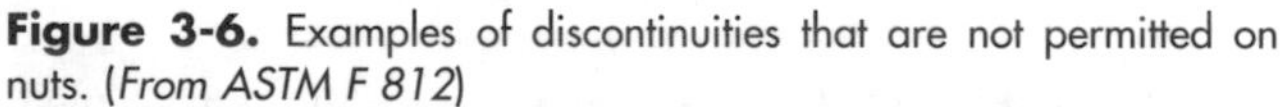

Figure 3-6. Examples of discontinuities that are not permitted on nuts. (*From ASTM F 812*)

are less than the minimum depth requirements of ASTM F812, "Specification for Surface Discontinuities of Nuts."

- Measurement. Check measurement complies with purchase order.

Supplementary Receiving Inspections

- Thickness. When applicable, measure thickness.

PIPE FITTINGS AND FLANGES

Applicable Standards. For pipe fittings and flanges ASTM standards are used to describe the material and ASME or API standards to describe the product type.

Minimum Information to Specify

- Manufacturer (if applicable)
- Manufacturer's part number (if applicable)
- NPS (e.g., NPS 2)
- Schedule (e.g., 40S)
- Rating (e.g., class 150)
- Material Standard (e.g., ASTM A403)
- Embedded class/grade (e.g., WP304)

	FLANGE	FITTING
• Type	e.g., weldneck	e.g., long radius
• ASME/API descriptor	e.g., B16.5	e.g., B16.5
• Flange/End	e.g., 125-250 AA	e.g., socket welding

Basic Receiving Inspection

- Sampling. Per guidance from engineering organization.
- Certification.
  - Check the box to verify that the part number exactly matches what was ordered.
  - If MTR was specified on the purchase order (or for ASME code-approved material), match heat number on parts to MTR.

  - Verify chemistry and mechanical properties meet specification requirements
- Appearance.
  - Flange finish. No radial scratches or other marks that cross the whole gasket contacting surface are permitted. ASME B16.5 shows allowable imperfections on flange faces. See Figure 3-7.
  - Weld bevels on flanges and fittings. Bevel geometry must be in accordance with ASME B16.25. No mechanical deformations or scratches are permitted

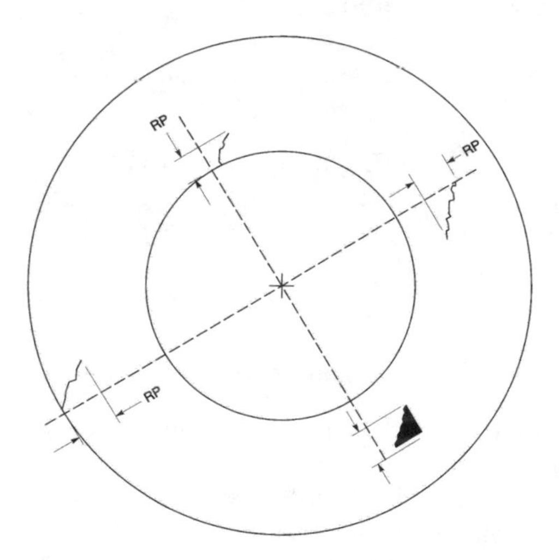

Figure 3-7. Amount of permissible scratching or damage on flange faces is strictly regulated.

NPS, in.	Maximum radial projection of imperfections that are no deeper than the bottom of the serrations, in.	Maximum depth and radial projection of imperfections that are deeper than the bottom of the serrations, in.
1/2	0.12	0.06
3/4	0.12	0.06
1	0.12	0.06
1-1/4	0.12	0.06
1-1/2	0.12	0.06
2	0.12	0.06
2-1/2	0.12	0.06
3	0.18	0.06
3-1/2	0.25	0.12
4	0.25	0.12
5	0.25	0.12
6	0.25	0.12
8	0.31	0.18
10	0.31	0.18
12	0.31	0.18
14	0.31	0.18
16	0.38	0.18
18	0.50	0.25
20	0.50	0.25
24	0.50	0.25

Notes: From ANSI/ASME B16.5

Imperfections must be separated by at least four times the permissible radial projection.

Protrusions above the serrations are not permitted.

Figure 3-7. Continued.

on the land or where the bevel ties into the OD of the fitting or flange. Minor scratches may be permitted on the bevel face it they are confined to face and will be completely melted out by the heat of welding.
- Concentricity. Verify that pipe fittings and flanges meet the specification (e.g., not honed or machined off center).
- Markings. Verify that markings are in accordance with applicable standard indicated on the purchase order.
- Coating. When applicable, check that the coating, such as galvanizing, rust preventative, or paint, meets the purchase order description and is not mechanically damaged.

Supplementary Receiving Inspections. The following additional inspections may be necessary to meet specific requirements of the purchase order for critical applications:

- Stub End Flanges. Specifically check for the following:
  - Length
  - Face fully welded (no circumferential gap is revealed by machining). See Figure 3-8.
- Liquid Penetrant or Magnetic Particle Examination. Check for cracks.
- Field Identification. For example, x-ray fluorescence or moly check to differentiate 316 stainless steel from 304 stainless steel.

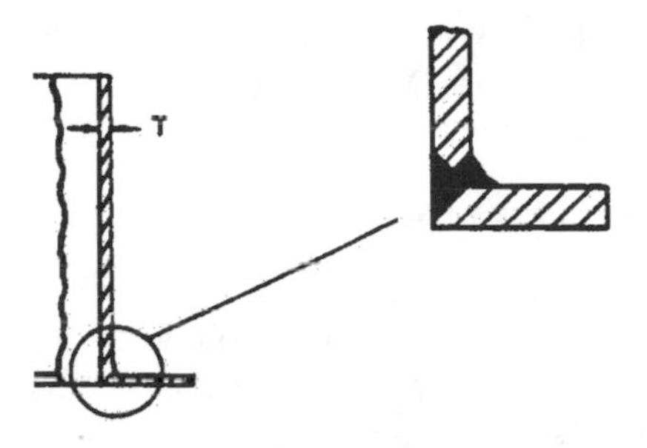

Figure 3-8. Face of stub end flanges must be fully welded with no weld discontinuities, such as lack of fusion, evident.

PIPE AND TUBING

Applicable Standards. Pipe and tubing material is principally covered by ASTM or API standards. Problems may occur when ASTM A53 grade F pipe is specified and received by the site. It is generic pipe, suitable for benign services or handrails. Unfortunately, it may become intermixed with more critical service pipe because grade F, like other grades of A53, it is not identified with surface markings for product below 2 inches in diameter.

Minimum Information To Specify. The following information is required to adequately specify pipe and tubing:

- ASTM or API standard
- Manufacturer

- NPS (e.g., NPS2)
- Schedule (e.g., 40S)
- Length

Basic Receiving Inspections

- Sampling. Per guidance from engineering organization.
- Certification. Verify that the following items match the product description on the purchase order:
  - ASTM specification (e.g., A312)
  - Embedded grade within specification (e.g., 316L)
  - NPS (nominal pipe size) (e.g., NPS2)
  - Schedule (e.g., 40S)
- Appearance. Visually inspect for the following:
  - No difference in finish, color, shape and size between lengths. Stainless steel pipe should contain no surface residues (e.g., tape adhesive). This requirement is not spelt out in industry standards and the user must specify it if required.
  - Straightness (e.g., check by rolling pipe on a flat table).
  - Markings in accordance with purchase order, as indicated in applicable standard, including manufacturer and heat number when applicable
  - Coating is not excessively damaged (e.g., galvanizing, rust preventative, paint).
  - Thread protectors furnished on threaded pipe. This is not necessarily specified in industry standards and must be indicated by the user if required.

- Measurement
  - Length in accordance with purchase order
  - Correct NPS and schedule

Supplementary Receiving Inspections

- Field Identification. Supplemental chemistry verification by x-ray fluoresence spectroscopy or chemical spot test (e.g., for moly in 316 stainless steel) may be required.

PLATE, SHEET AND STRIP

Applicable Standards. Plate, sheet and strip made to ASTM or ASME standards

Minimum Information to Specify. The following information is required to adequately specify plate, sheet and strip:

- ASTM or ASME specification
- Embedded grade within specification
- Length and width
- Thickness (gauge)

Basic Receiving Inspections

- Usually each piece is inspected.
- Certification. Verify that the following items match the product description on the purchase order:

  - ASTM or ASME specification
  - Embedded grade within specification
  - Thickness
  - Length and width
  - Special packaging requirements
- Appearance. Verify appearance for the following:
  - No surface damage, such as laps, dents or dings.
  - Flatness per specification
  - Markings in accordance with purchase order
  - Coating (if applicable) not damaged in any way, such as galvanizing or paint
- Measurement
  - Length and width of each piece in accordance with purchase order
  - Correct thickness per purchase order

Supplementary Receiving Inspections

- Field identification. Supplementary chemistry verification by x-ray fluorescence spectroscopy or chemical spot test (e.g., moly for 316 stainless steel) may be required.

VALVES

Minimum Information to Specify. Valves are usually specified by the end users internal designation number that cross references approved manufacturer's number and size ranges. Basic information required to fully identify a valve is as follows:

- End user's internal designation number
- Pressure rating
- Temperature rating
- Valve type (e.g., ball, globe, gate)
- Size
- End (e.g., socket, flange, bevel)
- Seat
- Seals
- Stem
- Stem packing
- Bolts
- Nuts
- Special design features
- Ball or disc
- Vendor
- Vendor's code for valve

Basic Receiving Inspections

- Sampling. Per guidance from engineering organization.
- Certification. Verify that the following items match the product description on the purchase requisition:
  - Certificate of compliance indicates valve meets specifier's internal designation for valve and manufacturer's number
- Appearance. Visually inspect for the following:
  - Manufacturer's tag is attached to valve and corresponds to specifier's internal designation. Note that nameplates or tags do not usually apply to brass or

bronze valves under 2 inches. These are considered commodity components.
- Pressure class and body material match purchase requisition.
- Valve openings are covered or suitably protected.
- Hand wheels, lever or chain wheels are securely attached and are of appropriate type for valve size.
- Internal and external surfaces of castings meet requirements of MSS SP-55. For internal surfaces, pull the caps and look down bores with dental mirror for evidence of damage, such as heavy machine marks, deep pitting or cracks.
- Flange finish. No radial scratches or other marks that cross the whole gasket contacting surface are permitted. ASME B16.5 shows allowable imperfections on flange faces. See Figure 3-7.
- Weld bevels on flanges and fittings. Bevel geometry must be in accordance with ASME B16.25. No mechanical deformations or scratches are permitted on the land or where the bevel ties into the OD of the fitting or flange. Minor scratches may be permitted on the bevel face it they are confined to face and will be completely melted out by the heat of welding.

• Measurement
  - Size matches purchase order description.
  - End to end dimensions meet purchase order description.

Supplementary Receiving Inspections

- Field Identification. Supplemental chemistry verification of valve body by x-ray fluorescence spectroscopy or chemical spot test (e.g., for moly in 316 stainless steel) may be required. The stem is usually too narrow to be identified by x-ray fluorescence spectroscopy and supplementary means of analysis may be required.
- Specified Cleanliness. Check when called out for special services, such as oxygen or chlorine.

GASKETS

Applicable Standards. Cut gaskets and envelope gaskets are dictated by ASME B16.2, "Nonmetallic Flat Gaskets for Pipe Flanges" and made to manufacturer's standard designations. Spiral wound gaskets are dictated by ASME B16.21, "Metallic Gaskets for Pipe Flanges—Ring Joint, Spiral-Wound and Jacketed" and made to manufacturer's standard designations.

Minimum Information to Specify

- Internal code designation
- Flange size
- Flange rating
- Style (e.g., sheet, spiral wound gasket)
- Thickness
- Flange standard
- Quantity

Basic Receiving Inspections

- Sampling. Per guidance of engineering organization.
- Certification. Check that product description or part number matches purchase order. For spiral wound gaskets, verify that the manufacturer's stamped or inscribed name, pressure class, size and materials are in accordance with the appropriate ASTM product-marking requirements.
- Appearance. For cut and envelope gaskets verify there are no cuts, damage, cracking, discoloration, or warpage; check there is no tape or other product that could affect sealing capability. For spiral wound gaskets ensure no dents and damage and that winding tack welds are secure. Verify color coding to indicate material of inner ring (when gasket has centering ring). The color coding is on the outer ring. See Figure 3-9.
- Measurement
  - Measure ID and OD using a ruler for conformance with purchase order. For sizes greater than 12 inches, take the average of two diameters at right angles to one another.
  - Measure thickness to ensure conformance with purchase order. For sheet/envelope gaskets or spiral wound gaskets, measurements must be in conformance with ASME B16.21 and B 16.20, respectively.

Suplementary Receiving Inspections. The following additional inspections may be necessary:

- For envelope gaskets, full face and reduced face dimensions meet designation requirements.

Color	Filler
No stripe	Asbestos
Pink	Mineral /graphite
Gray	Flexible graphite
White	"Teflon"
	Metal Winding
Yellow	304 sst
Green	316 sst
Orange	Monel
Red	Nickel
Beige	Alloy C-276 (Hastelloy*)
Gold	Alloy 600 (Inconel*)
No designated color	Other materials

* Gaskets without centering rings are not color coded.

Figure 3-9. Color of imprint on spiral wound gasket outer ring indicates type of alloy or material.

O-RINGS

Applicable Standards. O-ring sizes are specified according to AMS A568A, "Uniform Size Numbering System." O-ring materials are specified per manufacturer's and specifier's internal designations.

Minimum Information to Speficy. O-ring specification includes the following:

- Specifier's internal designation
- Manufacturer's compound code
- Polymer family (e.g., Viton GLT)
- Hardness (e.g., Shore A 75)
- Size, A568A dash number (e.g., -014)

Basic Receiving Inspections

- Sampling. Per guidance of the engineering organization.
- Certification. Verify that product matches packaging requirements. Maintain packaging if o-ring has no stencilling.
- Appearance. The following items may be checked:
  - Color coding. Reconcile with manufacturer's color coding system. Since carbon black is the optimum filler material, color coding may not be an issue.
  - Workmanship. Visually inspect for discoloration, flashing (parting line of mold is barely visible or invisible), and cracking (none permitted).

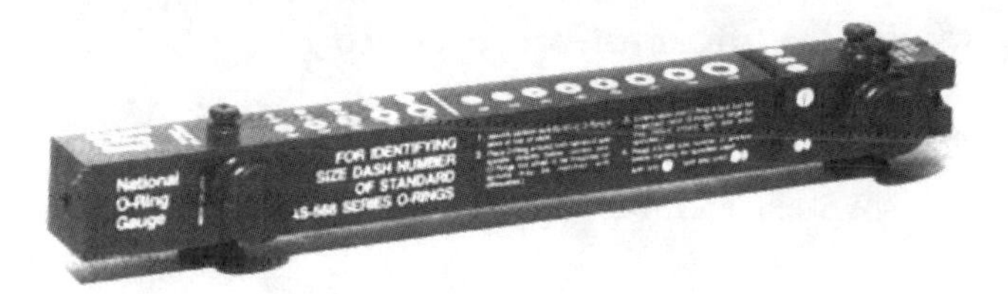

Figure 3-10. O-ring diameter may be measured on an o-ring gauge.

- Measurement
  - Diameter per the standard o-ring chart or standard o-ring gauge. See Figure 3-10.
  - Thickness. Check thickness on standard o-ring gage.

Supplementary Receiving Inspections

- Not usually applicable.

WELDING, FILLER METALS

Applicable Standards. The following standards are applicable to specifying welding filler metals:

- American Welding Society (AWS) A5.01, "Filler Metal Procurement Guidelines."
- American Society of Mechanical Engineers (ASME) Boiler and Pressure Vessel Code Section II Part C, "Specifications for Welding Rods, Electrodes and Filler Metals."

Minimum Information to Specify

- AWS or ASME specification number (e.g., A5.18 or SFA-5.18)
- AWS embedded classification (e.g., 7018-1, ERNiCr-Fe-3, etc.)
- Diameter
- Length (for rods)
- Quantity or unit package type and weight (e.g., carton, spool, reel)

Basic Receiving Inspections

- Sampling. Weld filler metals packages or containers are usually 100% inspected, although the packages or containers must not be opened.
- Certification. Verify that the product description on the package matches the purchase order. Verify the materials test report (MTR) that accompanies the product matches by heat or lot number for the following:
  – AWS or ASME specification number
  – AWS embedded classification
- Workmanship. Ensure package or container is not damaged.

Supplementary Receiving Inspections. The following supplementary receiving inspections, if required, necessitate opening of the package or container to remove a sample:

- XRF Analysis of Wire. Note that coated rod cannot be analyzed because some of the alloying additions are contained in the coating. For critical service it may be necessary to make a test weld pad with one or more electrodes and analyze it.

RECEIVING AND VERIFYING MATERIALS SUPPLY CHAIN—ROLES AND RESPONSIBILITIES

The many handoffs in the materials procurement supply chain makes it difficult to control the quality and accuracy of items. Job function responsibilities at the receiving and verifying end of the supply chain are as follows:

Receiving (vendors, stores, mechanics, operators):

- Ensure all critical service items are inspected and verified.
- Understand types of inspections required for different commodities.
- Obtain appropriate training and follow-up on site for specifying and inspecting activities for critical materials.
- File nonconformance report where applicable.

User (mechanic, operator, construction craft person)

- Ensure all critical service items are inspected and verified.

- Ensure all components used for new installation, maintenance, repair or replacement meet requirements of the process.
- Know how to locate and use applicable standards and codes to get information.
- Select materials correctly from stores codes and computerized systems.
- Recognize effects of changing process conditions and shutdowns on effectiveness of critical materials in system.
- Report persistent failures that may lead to injuries, litigation, excessive downtime, and high maintenance cost so that qualified failure analysis may be carried out.
- File nonconformance reports when applicable.

4
Metals

Summary: This chapter describes metals commonly used in the chemical and hydrocarbon processing industries. It helps the specifier, procurer, receiver and field user to understand how substitutions my occur and when to flag potential problems with serious consequences.

Metals are the most important materials of construction and are available in thousands of types (alloys), in various product forms, made to different quality levels.

METALS VS. ALLOYS

Pure metals, which are strictly chemical elements, are soft, exhibit low strength and have limited engineering applications. Alloys are materials composed of two or more chemical elements, at least one of which is a metal. Alloying develops properties that pure metals cannot achieve, such as strength, corrosion resistance or hardness. The terms metals and alloys are used interchangably.

Metals are distinguished from nonmetals by their strength, toughness, electrical conductivity and thermal conductivity. However, the dominant property that causes metals to be preferred over most nonmetals is their ability to deform in the presence of excessive stress rather than fracture catastrophically. This is the prime reason for the widespread use of metals for structural applications such as plant equipment, piping and pressure vessels.

Metal families include steels, cast irons, stainless steels, and nickel, copper, aluminum and titanium alloys.

Substitutions between alloys must not be made without a formal management of change process.

STEELS

Steels are the most important structural metals.

Carbon steels consist of iron to which carbon and manganese are added. Carbon has the greatest influence on properties, contributing principally strength and hardness. See Figure 4-1.

Low carbon steels contain less than 0.3% carbon, are generally weldable without preheat or postheat, and provide the majority of products for piping, pressure vessels, storage tanks and structures. Examples include the following:

- 1018, 1020 and 1030, used as plate and bar
- ASTM A36 or ASME SA-36, used for general structural applications

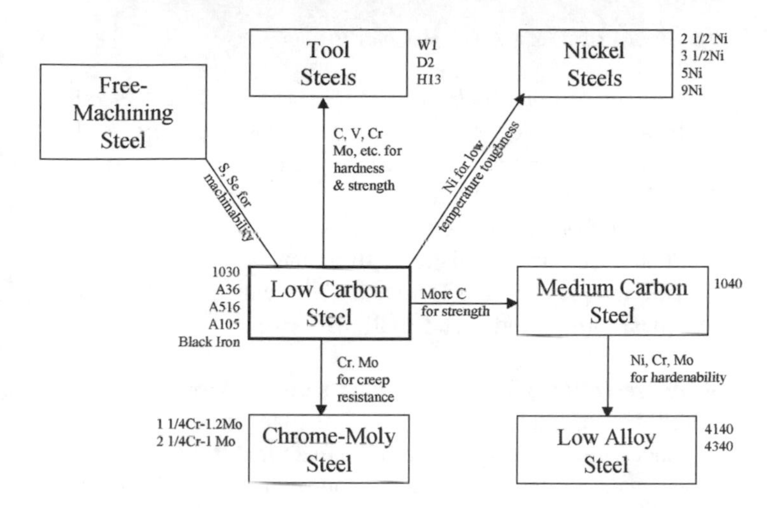

Figure 4-1. Steels are the largest used group of materials.

- ASTM A285 grade C, used for storage tanks, or ASME SA-285 grade C, used for pressure vessel plate
- ASTM A516 grade 70 or ASME SA-516 grade 70, used for pressure vessel plate with finer grain size for improved toughness.

Grain size is an important property of low carbon steels. Small (fine) grain size contributes to improved toughness, particularly at lower temperatures. Finer grain size is obtained by special steelmaking practices, as required in ASTM A516 grade 70/ASME SA-516 grade 70 (above). Never make an arbitrary change in a pressure vessel steel without consulting engineering.

Medium carbon steels are stronger and must be welded with care. Examples include 1040 and 1050. They are used for machinery components such as shafts, connecting rods, crane hooks, gears and axles.

High carbon steels are very strong and hard, difficult to weld, and have wear resistant rather than structural applications. Examples include 1080 and 1090, which are used for punches, springs and high tensile wire.

Alloy steels (low alloy steels) contain up to 5% of elemental additions, in addition to carbon and manganese. Alloying enhances mechanical properties, machinability, abrasion resistance, hardenability, corrosion resistance, magnetic properties, etc. Alloy steels are favored over carbon steels for demanding applications in components where significantly higher strength and hardness are required. Examples of common alloy steels include:

- 4130, available in plate and bar has relatively good weldablility
- 4140, available in plate and bar and may be through hardened in thicknesses up to $1\frac{1}{2}$ to 2 inches
- 4340, available in plate and bar and may be through hardened in thicknesses up to 4 to 6 inches
- 8620, available in plate and bar is weldable and formable, and often used where case hardening by carburizing is required

Hardenability is a property that indicates how fully a steel responds to hardening during heat treatment. Hardenability must not to be confused with hardness. For example, 1040 steel and 4140 steel develop the same surface hardness on quenching because of similar carbon content (.4%). However, 4140 achieves this hardness value to a greater depth than does 1040, so that it develops better strength and toughness in thicker sections after tempering. 4340 has even better hardenability and is used in the most demanding applications. Low alloy steels are favored over medium carbon steels for mechanical components such as shafts and machinery when the section size exceeds 2 inches, but only when quenched and tempered. If you use an alloy steel that is not in the quenched and tempered condition (e.g., anncaled) you are wasting your money. See Figure 4-2.

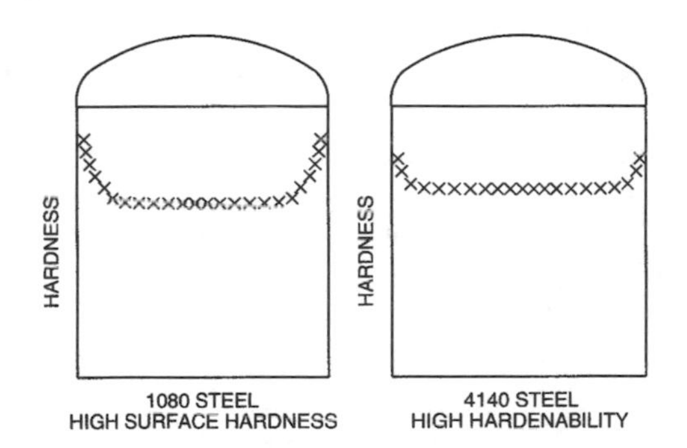

Figure 4-2. Hardenability is a measure of the depth to which a steel may be hardened on quenching.

Nickel steels contain 2% to 9% nickel to increase toughness at low temperatures from temperatures of 0°C to –195°C. Applications include plant piping and storage tanks for liquefied hydrocarbon gases.

Chrome-moly steels have $\frac{1}{2}$% to 9% chromium and either $\frac{1}{2}$% or 1% molybdenum. Chromium increases scaling resistance and molybdenum improves or provides elevated temperature strength. Carbon content is kept low to maintain weldability. Chrome-moly steels are widely used for piping and pressure vessels operating up to 1000°F in environments such as steam or hydrogen and are available in many product forms. The two most popular alloys are $1\frac{1}{4}$ chrome-$\frac{1}{2}$ moly and $2\frac{1}{4}$ chrome-1moly.

Tool steels are a diverse family with high carbon and high alloy contents. They are the strongest, hardest and most wear resistant steels, but lack toughness and weldability. Tool steels are designed for specific uses and are identified by a letter indicating the group followed by one or two numbers.

Group W is water hardening, essentially high carbon steels made to improved melting practices. Group W is used for cutlery, forging dies and hammers

Group S is shock resisting, for repeated impact loading. Group S is used for chisels, rivet seats and structural applications.

Groups O, A and D are cold work tool steels and are the most popular groups of tool steels. They have moderate-high wear resistance and moderate-poor toughness. They are used for dies, rolls and gear pumps (e.g., O1 and D2).

Group H covers hot work tool steels, which retain strength, toughness and resistance to softening at elevated temperatures. Group H steels are very strong and have important structural applications, such as rolls and fasteners (e.g., H11 and H13).

Groups M and T retain softening resistance to high temperatures and are used for specialized tooling applications.

Groups P and L are mold and special purpose tool steels, respectively, and have niche applications.

CAST IRONS

Cast irons are easily poured into complex shapes and have relatively low toughness, ductility and weldability. See Figure 4-3.

Gray iron has the greatest usage, with excellent damping capacity, but no ductility. It is produced in various grades identified by the tensile strength in ksi; for example grade 40, which is the most commonly used gray iron. Gray irons are used for bases and supports for moving components to dampen vibrations and for wear resistant liners in cylinder sleeves and valve guides.

White iron is formed by fast cooling the gray iron composition in the mold to produce an extremely hard structure for wear resistant items, such as heavy-duty machinery components.

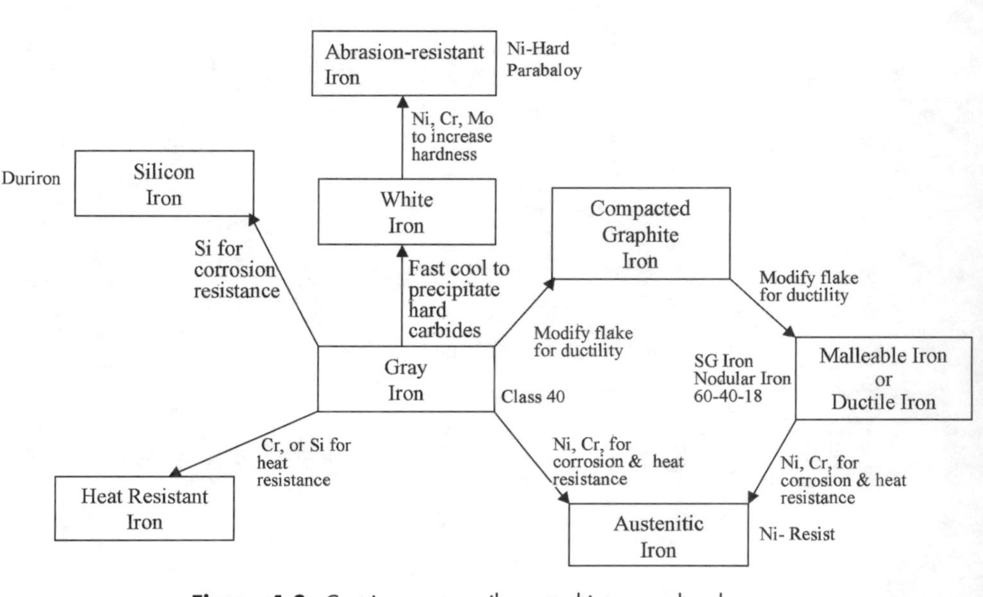

Figure 4-3. Cast irons are easily poured into complex shapes.

Malleable and ductile irons are cast irons modified by alloying and heat treatment to improve ductility. For example, grade 60-40-18 ductile iron has equivalent tensile strength to class 60 gray iron, with 40 ksi yield strength and 18% elongation. Malleable and ductile irons are used for low-pressure flanges and piping and various machine parts.

Compacted graphite iron is midway between gray and ductile or malleable iron in properties. It is primarily used in automotive applications.

Alloy irons contain up to 30% of various alloying elements, which are added to improve corrosion, abrasion or heat resistance. Alloy irons are better known by trade names. They find many niche uses, such as pump impellers, grinding mill liners, or burner nozzles, and are limited by their lack of weldability.

STAINLESS STEELS

Stainless steels are the most widely used alloys for heat and corrosion resistance and low temperature toughness. Wrought stainless steels are named for their metallurgical structure: martensitic, ferritic, austenitic, duplex, and precipitation hardening. Cast stainless steels groups are heat resistant and corrosion resistant.

Martensitic stainless steels have the least chromium and are the least corrosion resistant. They are quenched and tempered like carbon steels to increase strength and are strongly magnetic. The base alloy is 410 and they are used for turbine buckets, valves and wear resistant parts such as dies. See Figure 4-4.

Ferritic stainless steels contain more chromium than martensitic stainless steels to improve corrosion resistance. They cannot be hardened by quenching and tempering and have limited ductility in thicker sections. The base alloy is 430. They are used for heat exchanger tubes. See Figure 4-5.

Austenitic stainless steels develop a unique structure through the addition of nickel. They have the widest range of application of all corrosion resistant materials, with excellent corrosion resistance, weldability, high temperature strength and low temperature toughness. They are relatively weak compared with martensitic and precipitation hardening stainless steels. They are not hardened by quenching and may be strengthened only by cold work. The base alloy is 302 and the two most common grades are 304L and 316L. In addition to the standard grades many proprietary alloys have been developed with higher strength and better corrosion resistance. Austenitic stainless steels are used for instrumentation, piping, pipe fittings, fasteners, pressure vessels, storage tanks, etc. See Figure 4-6.

Dual marked stainless steels. In years past, the two most common austenitic stainless steels were available in two

(text continued on page 86)

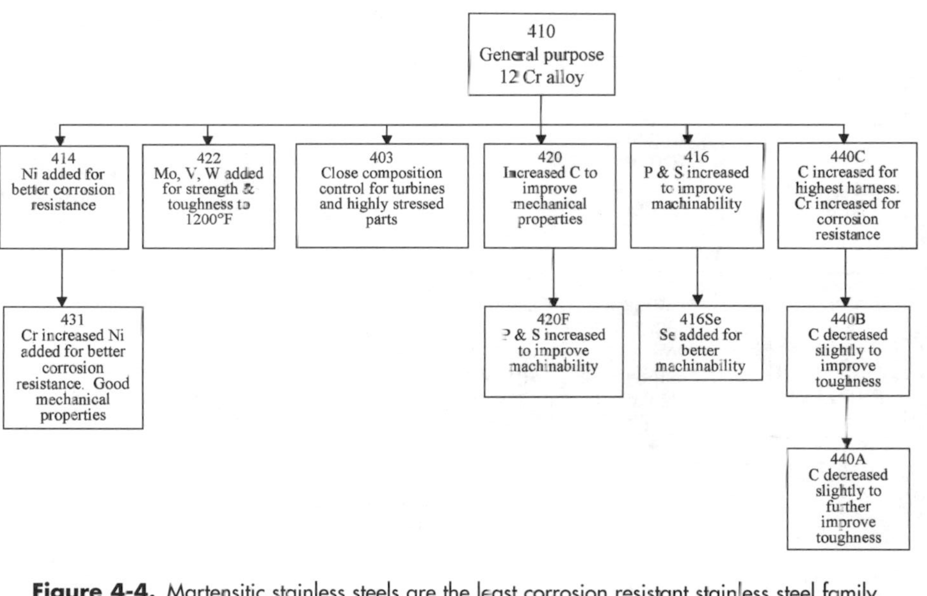

Figure 4-4. Martensitic stainless steels are the least corrosion resistant stainless steel family.

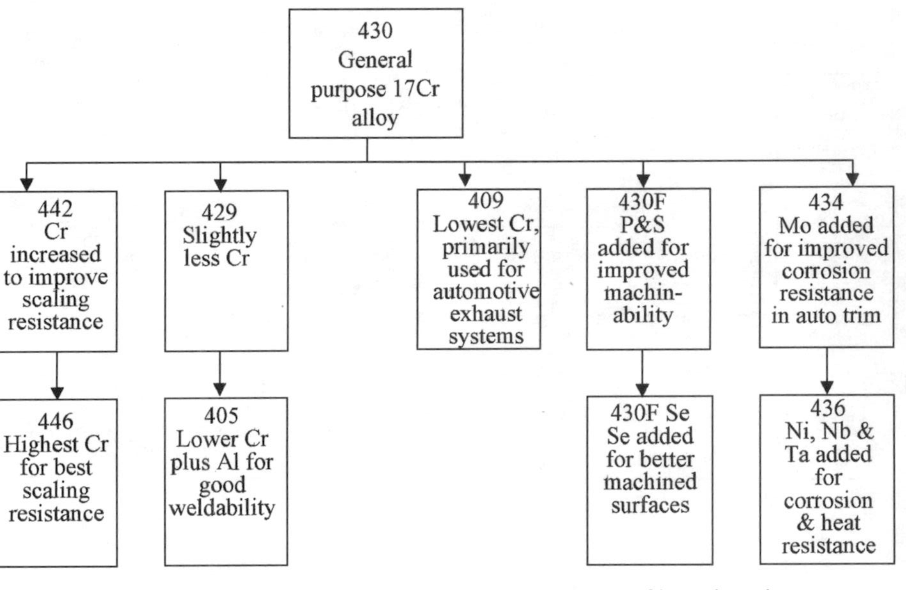

Figure 4-5. Ferritic stainless steels have limited use because of limited toughness.

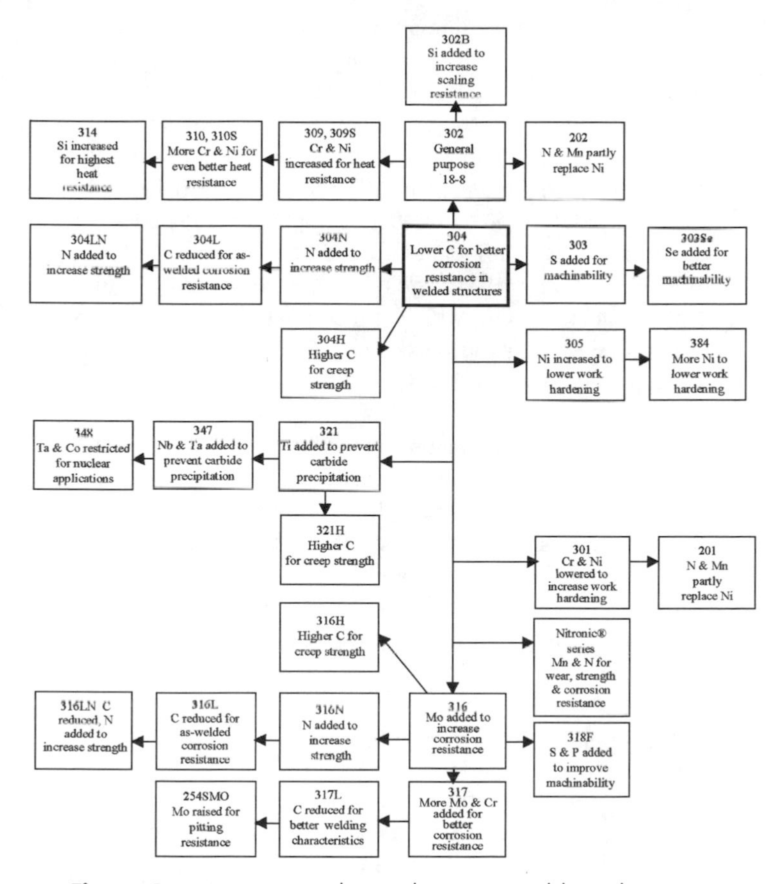

Figure 4-6. Austenitic stainless steels are most widely used corrosion resistant materials.

(text continued from page 82)

grades: those with regular carbon content ("straight grade") and those with extra low carbon ("L grade"). They were identified as 304 and 316 or 304L and 316L.

When welding the straight grades it was possible to "sensitize" the base metal adjacent to the weld. Sensitizing occurs when alloys with more than 0.03% carbon are held in the 1,200 to 1,900°F (650 to 1,040°C) temperature range. Excess carbon combines with some of the chromium in the alloy to form chromium carbides, leaving the surrounding region depleted in chromium, which is the alloying element chiefly responsible for corrosion resistance. Consequently, the alloy develops thin localized bands on either side of the weld that have significantly impaired corrosion resistance in specific environments. Sensitizing during welding is avoided by using L grades with carbon content restricted to less than 0.03%.

Improved stainless steel melting practices improved efficiency in carbon content reduction, so the L grade became cost competitive with the straight grade. Since the reduction of carbon is achieved at the expense of strength, producers began to add nitrogen to offset the loss. Thus dual marked grades are now commonplace (304/304L and 316/316L). They combine the carbon content of an L grade and the strength improvement of a straight grade.

Do not substitute an L grade when a straight grade is specified and dual marked product is not available. The higher strength of the straight grade may be an essential requirement of the design. This is particularly important with high temperature services where L grades are substantially weaker than straight grades.

Duplex stainless steels are composite materials with a structure of approximately equal amounts of austenite and ferrite. They are easier to fabricate than ferritic stainless steels and are tougher. They are twice as strong as austenitic stainless steels and more resistant to chloride stress cracking. However, they require considerably more care on welding and during fabrication. Most duplex stainless steels are proprietary alloys. They are used for heat exchanger tubing, pressure vessels and piping. See Figure 4-7.

Precipitation hardening stainless steels are alloyed to achieve significantly higher strengths than other stainless steels by heat treatment. They are purchased in the soft, annealed condition and machined or formed to final shape before being hardened. Various heat treatment temperatures are used to achieve different strength levels. For some grades the hardened condition is indicated by the letter H followed by the hardening temperature in degrees F (e.g., 17-4PH, H1100). Precipitation hardening stainless steels are used in critical high strength applications where resistance to fracture propagation is very important, such as valve stems, fasteners and shafting. See Figure 4-8.

> Precipitation hardening stainless steels must never be used in the as-purchased annealed condition. They are not only soft and weak, but they lack toughness and will fail prematurely.

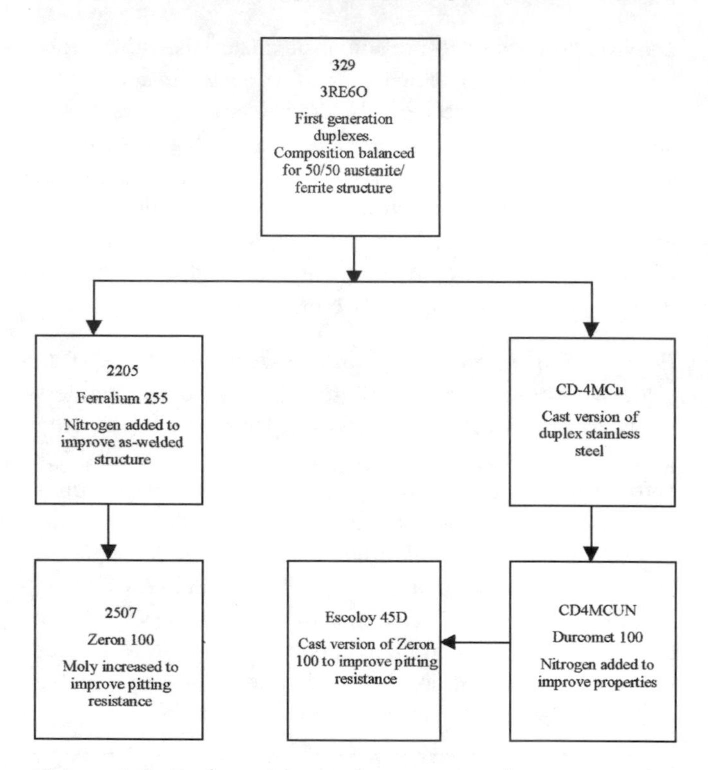

Figure 4-7. Duplex stainless steels are stronger than austenitics but require care during welding and fabrication.

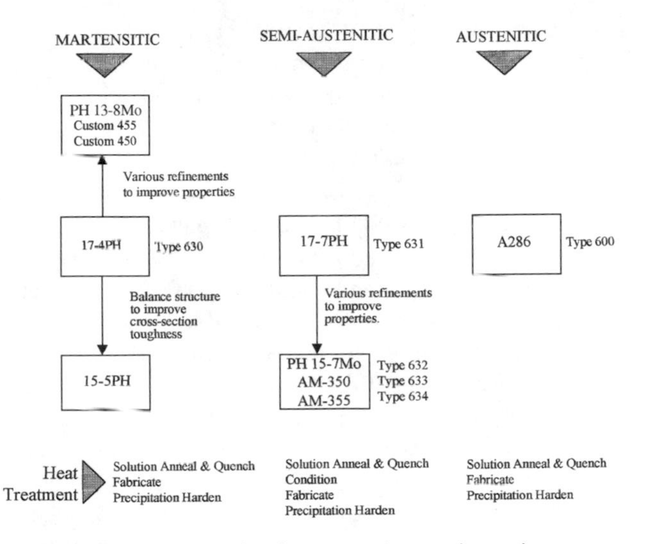

Figure 4-8. Precipitation hardening stainless steels are the strongest group of stainless steels.

Corrosion resistant castings are stainless steels and nickel alloys with various metallurgical structures. They are identified by the prefix C followed by a letter indicating the approximate alloy content. The higher the letter (A being lowest) the greater the alloy content. A number and letter following a dash indicate the carbon content and any important alloying elements. The most common corrosion resistant casting is CF-3M, equivalent to 316L wrought stainless steel. Corrosion resistant castings are used for valves, pumps and centrifuge parts where it is not possible or economical to fabricate an equivalent wrought alloy. Manufacturers may use their internal descriptions to specify proprietary alloys. See Figure 4-9.

> Take care when specifying cast pump and valve corrosion resistant castings to match the materials designations of wrought piping system components. Corrosion resistant castings are generally available in more limited grades. For example, although wrought 304L and 316L are widely available, the cast version of 304L (CF-3) is much less available. Foundries prefer to produce the cast version of 316L (CF-3M) because it has much broader corrosion resistance and can usually be substituted for CF-3 without problem. Consequently, specifying CF-3 valves and pumps in a 304L piping system may incur price and delivery penalties.

Heat resistant castings are stainless steel and nickel alloys with various metallurgical structures. They are identified by the prefix H followed by a letter indicating the approximate alloy content. The higher the letter the greater the alloy content. Heat resistant castings have high creep strength and are used for reformer and pyrolysis tubing and related parts. Although most heat resistant alloys use generic designations, there are many proprietary compositions based on type HP. See Figure 4-10.

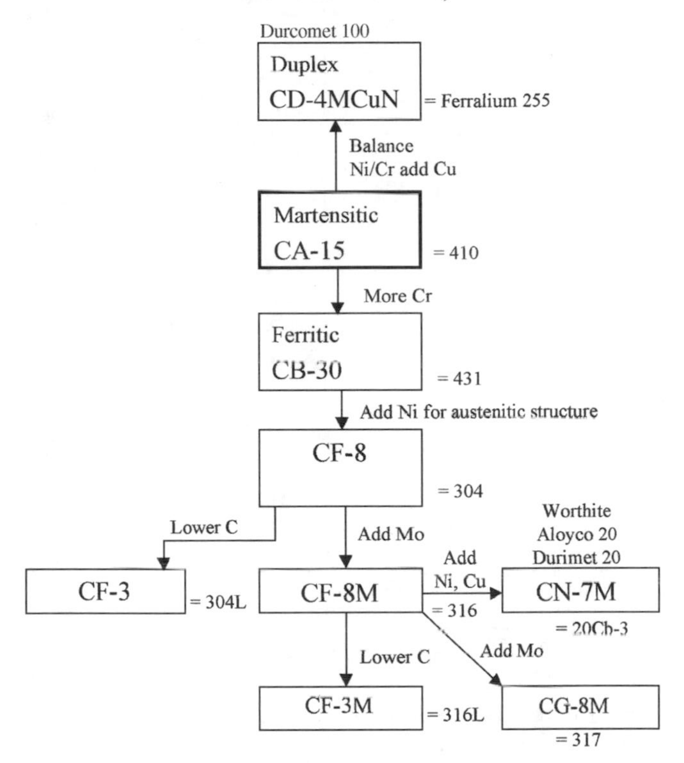

Figure 4-9. Corrosion resistant castings comprise stainless and nickel alloys having several various metallurgical structures.

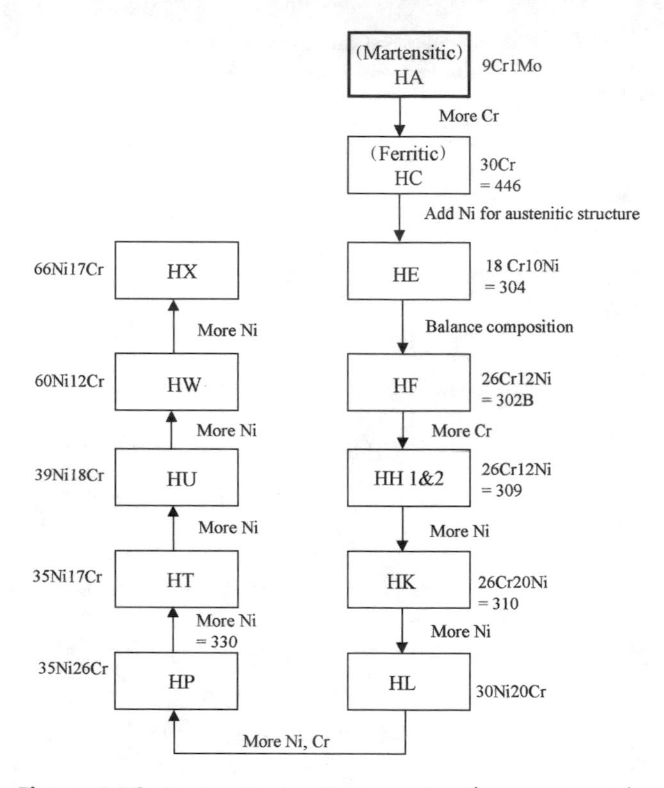

Figure 4-10. Heat resistant castings are primarily austenitic stainless and nickel alloy materials.

NICKEL ALLOYS

Nickel has many uses and is incorporated as a major or minor constituent in approximately 3,000 alloys. Nickel alloy groups are based on nickel and nickel-chromium. There are many proprietary nickel and nickel-chromium alloys. See Figure 4-11.

High nickel alloys are essentially pure nickel with strict control of elements such as carbon. For example, nickel 200 is used for high temperature concentrated caustic, in vessels, coils and pump components.

Nickel-copper alloys have good corrosion resistance. For example, Monel 400 is used in many applications, including salt water or brackish services, for pumps, tubing and piping.

Nickel-iron alloys have specific physical properties, such as magnetic response or low thermal expansion, giving them niche applications (e.g., for glass to metal seals and linkages that require little or no thermal expansion).

Nickel-moly alloys are extremely resistant to very specific chemical environments such as hot hydrochloric acid. Examples are the Hastelloy® alloy B series, which are used for vessels, columns and piping.

Nickel based superalloys are one of three groups of materials used for demanding, high-temperature applications such as jet engines and gas turbines (e.g., Inconel® X-750). The other two groups are iron-based and cobalt-based superalloys.

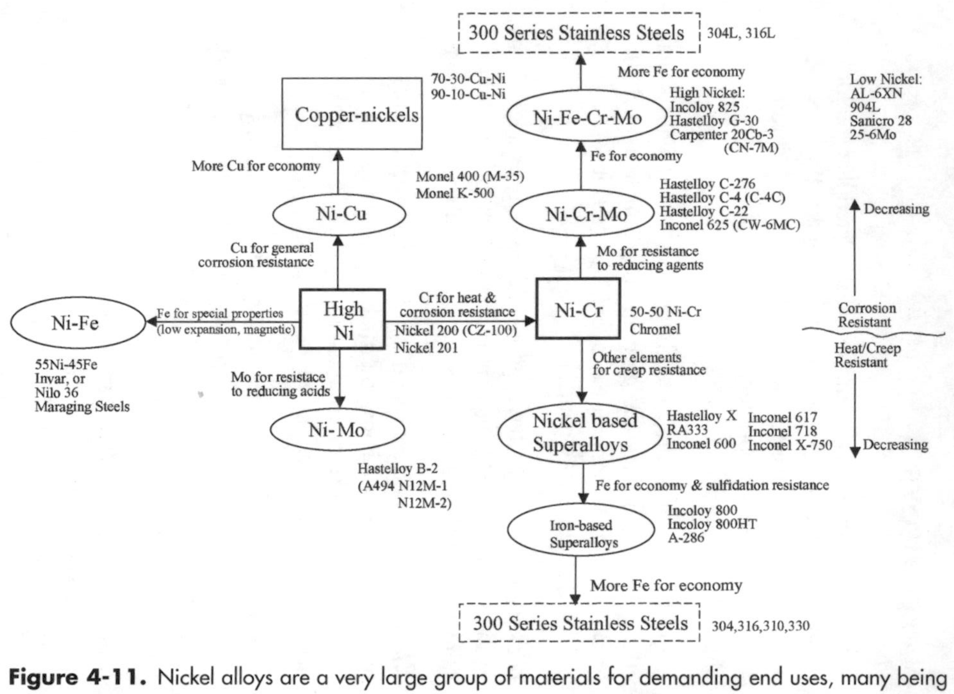

Figure 4-11. Nickel alloys are a very large group of materials for demanding end uses, many being proprietary.

Nickel-iron-chrome alloys include the iron base superalloys and the heat resistant castings (see previous section). Iron-base superalloys are an extension of the austenitic stainless steels and include alloys such as the Incoloy® 800 series. They are used for piping and manifolds in refining, power generation and petrochemicals.

Nickel-chrome-moly alloys have the best corrosion resistance of all the groups and are used for the most demanding corrosives for many applications, such as vessels, piping and instrumentation. Examples include the Hastelloy® alloy C series.

Nickel-iron-chrome-moly alloys are less costly than nickel-chrome-moly alloys. They are divided into the low nickel and high nickel groups. Alloys in the low nickel group

> What's in a letter? With nickel alloys it is important to make no mistake in entering the specification. For example, Hastelloy® C-276 is used in many demanding chemical environments. Hastelloy® B-2 also possesses exceptional chemical resistance, but to niche environments like hot hydrochloric acid. It fails rapidly if substituted for Hastelloy® C-276 in a service like ferric chloride. Never call out a material merely as "Hastelloy," "Inconel," etc. It is best to identify any nickel alloy by its UNS designation, such as UNS N102756 for Hastelloy® C-276 and N10665 for Hastelloy® B-2.

are sometimes called superstainless steels (e.g., 904L). Alloys in the high nickel group are based on Carpenter 20Cb-3, developed for sulfuric acid, and are sometimes called alloy 20 types. Both groups are used for their excellent corrosion resistance.

COPPER ALLOYS

Copper alloys consist of several families that have high electrical and thermal conductivity and corrosion resistance. See Figure 4-12.

Commercially pure coppers contain at least 99.9% copper and are used principally for their high electrical conductivity (e.g., in electrical components).

Modified coppers are commercially pure coppers with very small amounts of alloying elements for improvement in properties such as machinability or elevated temperature strength while maintaining high electrical conductivity.

Beryllium coppers contain small amounts of beryllium and are precipitation hardened to very high strengths for springs, switch components, retaining rings, etc.

Brasses are wrought copper alloys containing zinc as their principal alloying element. They are the most popular and

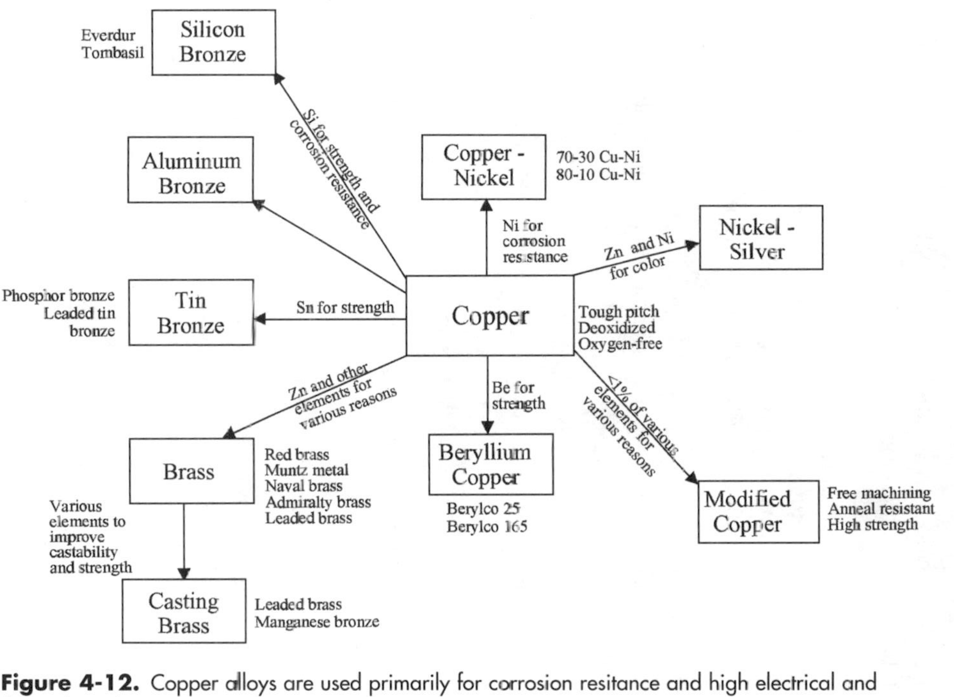

Figure 4-12. Copper alloys are used primarily for corrosion resitance and high electrical and thermal conductivity.

least expensive copper alloys and are used for condenser tubing, valve stems, etc.

Casting brasses contain additional alloying elements to improve castability and may be poured into complex shapes with low porosity and good mechanical properties. Applications include low pressure valves and fittings.

Tin bronzes are alloyed with tin and are also called phosphor bronzes. They have high strength, corrosion resistance, toughness and low friction and are suitable for bearings operating under high load.

Aluminum bronzes are alloyed principally with aluminum for useful strength and excellent corrosion resistance. Applications include valves, pumps and bushings.

Silicon bronzes are alloyed principally with silicon to achieve high strength, similar to carbon steel, coupled with good toughness and excellent corrosion resistance. Applications include fasteners, pumps and bearings.

Copper-nickels contain up to 30% nickel and have moderate strength and better corrosion resistance than other copper alloys. They are used for condenser tubing, tube sheets, salt water piping and ferrules.

Nickel-silvers contain zinc and copper and have a characteristic silvery color and provide an excellent base for chromium or silver plating. They have many uses, including hardware and parts for cameras and musical instruments.

ALUMINUM ALLOYS

Aluminum alloys consist of several families, used for low density (lightness), corrosion resistance and high strength-to-weight ratio. Aluminum alloy descriptions comprise the alloy designation plus the thermal-mechanical condition (temper designation), such as 6061-T6. See Figure 4-13.

1000 series alloys have low strength and are used for foil, decorative trim, low presure piping and storage tanks.

2000 series alloys contain aluminum and copper and are strengthened by precipitation hardening, although the resulting ductility and weldability are low. They have poor corrosion resistance and are not used significantly in process industry applications.

3000 series alloys contain manganese and are strengthened by cold work. These alloys and their cast equivalents have many applications where their moderate strength and corrosion resistance is an advantage, such as storage tanks, silos, cryogenic vessels and control valve bodies.

4000 series alloys contain silicon and are mainly used in architectural applications.

5000 series alloys contain magnesium and have reasonably high strength and good weldability. They are used for marine and other applications.

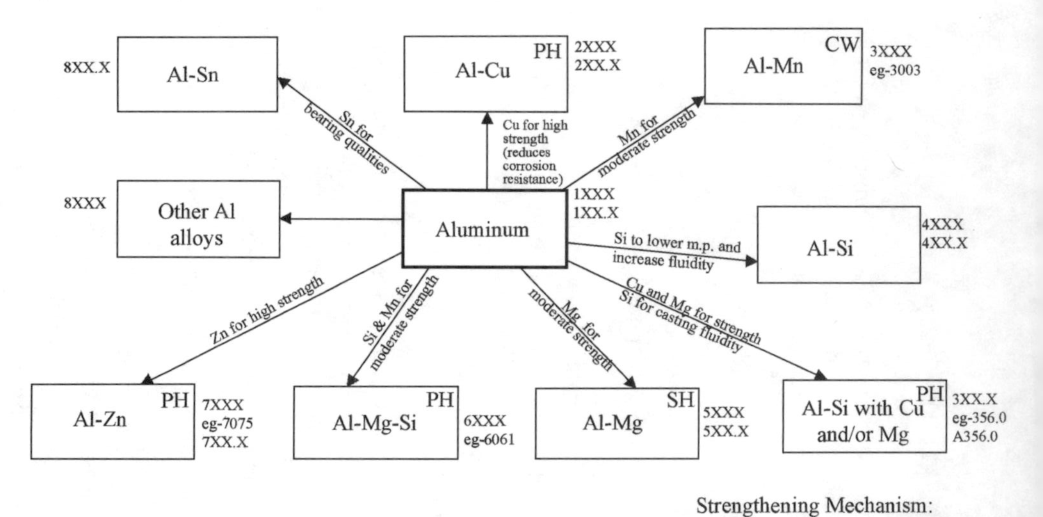

Strengthening Mechanism:

PH precipitation hardening

SH solution hardening

CW cold working

Figure 4-13. Aluminum alloys are used for low density and high strength-weight ratio.

6000 series alloys contain aluminum and silicon and are strengthened by precipitation hardening and are very popular. For example, 6061 is one of the most versatile aluminum alloys for transportation equipment, cryogenic vessels and welded structures.

7000 series alloys contain aluminum and zinc and are precipitation hardened to high strength levels. For example, 7075 is one of the strongest aluminum alloys available and is used for airframe structures and other highly stressed components.

8000 series alloys comprise alloys developed for high load bearings and bushings.

TITANIUM ALLOYS

Titanium alloys comprise low- and high-strength materials with low density and high strength-weight ratio. The process industry uses a limited number of titanium alloys chiefly for their corrosion resistance. See Figure 4-14.

Commercially pure titanium consists of several grades with increasing amount of oxygen and increase in strength. The versions most used are grades 2 and 3 for tubing, vessels and pipe.

Titanium-palladium alloys (grades 7 and 11) have significantly improved corrosion resistance and are used where grades 2 and 3 are inadequate.

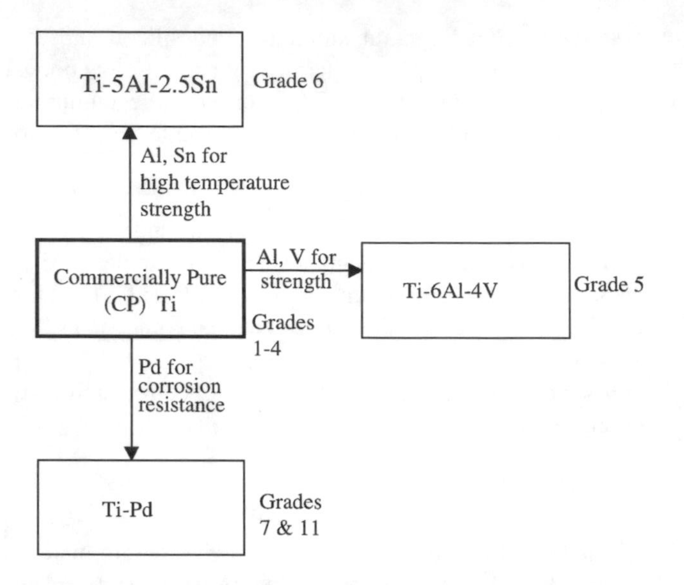

Figure 4-14. Titanium alloys are used for low density, high strength-weight ratio and excellent corrosion resistance.

Ti-6 aluminum-4 vanadium (grade 5) is the most widely used alloy for pumps, shafts, valves and high strength parts.

Ti-5 aluminum-2.5 tin (grade 6) is used for high temperature strength in process equipment and also as hardware to handle cryogenic liquified gases.

5
Metal Product Forms

Summary: This chapter describes metal manufacturing and heat treatment methods and their available geometries (shapes). Manufacturing method and heat treatment exert a significant effect on properties and service performance. Shape influences final fabrication cost. It is necessary to procure materials per the design requirement or drawing and verify the correct item has been received.

Metal products are made by distinct manufacturing routes, leading to specific product forms to suit end-user needs. Fundamental differences between wrought, cast, powder metal and weld filler metals have led to different specification systems for such products. Heat treatment is the most important method of strengthening metals although mechanical work is also used. See Figure 5-1.

Material Form	Characteristics	Specification System
Wrought	Bar tubing/pipe plate, sheet, strip Heat treatment is critical Welding may alter properties and suitability for service	ASTM, ASME, SAE, AMS
Cast	Produced to final shape Less homogeneous than wrought May not match wrought equivalent compositionally	ASTM, ASME, ACI, UNS, SAE
Filler Metal	Covered, bare wire/rod or flux cored forms May or may not match wrought equivalent compositionally	AWS
Powder Metallurgy	Compositionally homogeneous Porosity an advantage, e.g., for self lubricating bearings	MPIF

Figure 5-1. Wrought, cast, filler metals and powder metallurgy products have different characteristics, resulting in different specification systems for them.

WROUGHT METALS

Wrought metals comprise the majority of materials used in manufacturing plants. Wrought metals are initially cast into ingots or billets and then converted by mechanical working and forming processes into distinct product forms that may be fabricated into end products.

Mechanical working is classified as hot work and cold work. See Figure 5-2.

Hot work is mechanical working that does not lead to hardening. The metal remains relatively soft and weak. Hot working methods include hot rolling, hot forging and hot extrusion.

Cold work is mechanical working that causes hardening. The metal becomes stronger, harder and less ductile with increasing cold work. Cold working methods include cold rolling, cold drawing, and cold extrusion.

> Know whether hot worked or cold worked product is needed from the specification description and do not make substitutes.

Cold working and hot working are used to produce the following shapes:

Structural shapes are hot-rolled flanged shapes having at least one dimension of the cross section 3 inches or greater. They are used for construction applications such as beams.

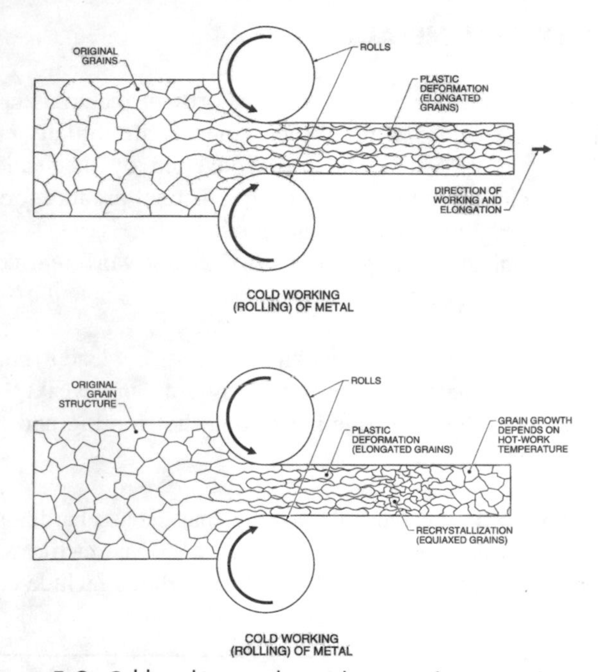

Figure 5-2. Cold working can be used to strengthen a metal, whereas with hot working the metal remains soft and ductile.

Bar is elongated hot or cold worked product, relatively thick and narrow, with a simple, uniform cross section that is rectangular, square, round, oval or hexagonal. Bar is produced in bar mills. Cold worked (cold finished) bar has better surface finish and dimensional accuracy than hot worked bar.

Wire is thin, flexible continuous product of circular cross section. Wire may be soft or cold drawn to high strength. Wire is made from wire rod which is drawn down to make wire. Wire rod is used to make fasteners

Tubing is hollow product of round, square or other cross section having a continuous perimeter. Tubing is designated in terms of outside diameter (OD), inside diameter (ID), wall thickness (schedule) or weight per foot, depending on product shape. Designations refer to the actual values or nominal (named) values.

Pipe is a hollow product of circular cross section made to standard combinations of OD and wall thickness. There are fewer sizes of pipe compared with tubing because pipe has standard sizing whereas tubing may be drawn to any dimension to suit the needs of the end-user. Pipe sizes are classified by their nominal pipe size (NPS), which has a value based on the pipe diameter. Up to and including 12 in. NPS, there are wide differences between NPS and OD. For example, $\frac{1}{2}$ in. NPS standard weight pipe has an actual OD of .840 in. and a theoretical ID of .622 in. When different weights per foot or wall thicknesses (schedules) are indicated for a particular NPS, the OD of the pipe is not changed, only the ID of the pipe is altered. For pipe sizes 14 in. NPS and greater, the actual OD of the pipe corresponds to the NPS.

Tubing or pipe may be seamless or welded. Seamless products are made from a billet, which is pierced through the center to make it hollow and then drawn to size. Welded products are made from skelp, a coil of flat strip that is first uncoiled and curved about its longitudinal axis into a circular configuration. The longitudinal edges of the skelp are welded together to form tubing using electric resistance welding (ERW) or filler metal. The weld bead may then be reduced in size (sized) by various methods such as scarfing, rolling, or drawing. See Figure 5-3.

> You must know the service category of heat exchanger tubing in order to specify it economically. There are three categories of heat exchanger tubing. For standard tubing, where leaks have little or no effect on the process or the equipment, the appropriate ASTM or ASME specification is applicable with no additional testing. For process tubing, where leaks will require plugging or replacement of the tubing, but not immediately, some additional testing will be required. For critical tubing, where leaks have a significant effect on the equipment, process, safety or the environment, the appropriate additional testing is required. See Figure 5-4.

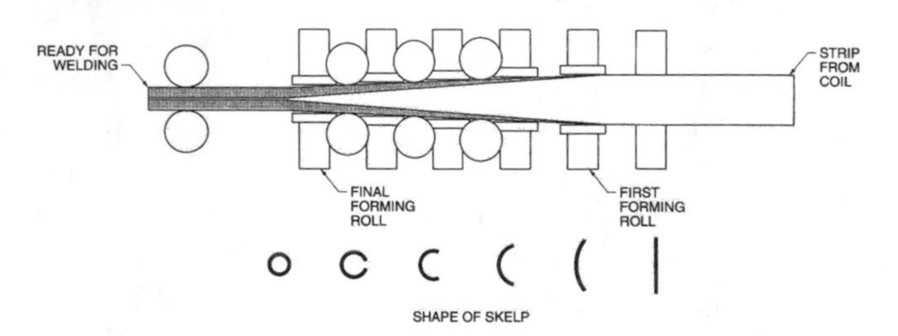

Figure 5-3. Welded tubing is made from strip, also known as skelp.

TUBING DESCRIPTION	HOW MADE	USES	STANDARD CATEGORY	PROCESS CATEGORY	CRITICAL CATEGORY
Welded	Weld bead on ID not removed	Basic	Minimum per applicable standard	Not normally used	Not normally used
Welded and heat treated	Weld bead on ID not removed	Basic	Minimum per applicable standard	Not normally used	Not normally used
Welded, sized, heat treated	Weld is flush on OD and ID by cold working	Many general process applications	Minimum per applicable standard	Eddy current and air under water	Not normally used
Welded, bead formed, heat treated	Weld is flush on OD and ID by cold working	Many general process applications	Minimum per applicable standard	Eddy current and air under water	Not normally used
Welded, 15% cold worked, heat treated	Weld is flush on OD and ID and tube wall reduced 15% minimum	For services that are aggressive to welds	Minimum per applicable standard	Air under water	Eddy current and air under water
Seamless, cold worked, heat treated	Pierced or extruded and then drawn to size	Critical applications	Minimum per applicable standard	Air under water	Eddy current and air under water

Figure 5-4. Required quality assurance testing level for heat exchanger or condenser tubing depends on the service category: standard, process, or critical.

Flanges and pipe fittings may be produced by forging processes such as ring rolling or impression die forging with the advantage that the metal grain flow is oriented to the contour of the part, giving highest strength in the direction of greatest stress. Less critical flanges and pipe fittings are also produced by other methods such as casting.

CASTINGS

Castings are made by pouring molten metal into a mold cavity of the desired shape and allowing the metal to solidify. Castings are advantageous when it is too expensive or difficult to create the same shape by metal forming, machining, or welding. Cast products include valve bodies, pump and compressor casings and certain types of flanges.

Selection of casting processes is determined by the required soundness (freedom from porosity), surface finish and desired unit cost for the part. See Figure 5-5.

Castings are distinctly different from wrought metals, for the following reasons:

- Castings are less homogeneous (compositionally uniform) because they are not hot worked to accelerate diffusion of chemical elements and require longer soaking times to achieve homogenization.
- Castings often exhibit slight compositional differences from equivalent wrought metals to improve castability.
- Castings contain high residual stress unless removed by annealing heat treatment.

Casting Type	Attributes
Sand	Versatile, most used
Centrifugal	For tubular products and parts of circular cross section with low porosity
Investment	For small, complex components with good surface finish and dimensional tolerances
Permanent Mold	For high production runs, with excellent soundness, surface finish and dimensional tolerance
Die	For complex shapes with thin walls in low melting point metals such as aluminum and zinc

Figure 5-5. Selection of casting process is determined by the required end properties and quality required.

- Castings do not exhibit anisotropy (directional properties) like cold worked metals

Because castings are fundamentally different from wrought metals it is important to indicate all needed requirements before they are specified for critical applications. See Figure 5-6.

Designers employ different principles from those used for wrought metals to take advantage of the unique characteristics of castings and ensure the molten metal solidification sequence in the design does not create cavities or porosity, or develop cracking from excessive stresses. See Figure 5-7.

✓	Item
	Product Standard (e.g., ASTM)
	Designation within Standard
	UNS Designation
	Pattern Number
	Drawing Number
	Corrosion Evaluation
	Special Nondestructive Examination
	Repairs Permitted or Not
	Certification Requirements
	Special Heat Treatment
	Identification Requirements
	Special Packaging

Figure 5-6. Castings require different sets of information for their specification compared with wrought metals.

POWDER METALLURGY

Powder metallurgy (P/M) is the forming of precision shapes and components from metal or nonmetal powders, or mixtures of the two. Components can be impregnated with oil or plastic, heat treated, plated, machined or forged. P/M

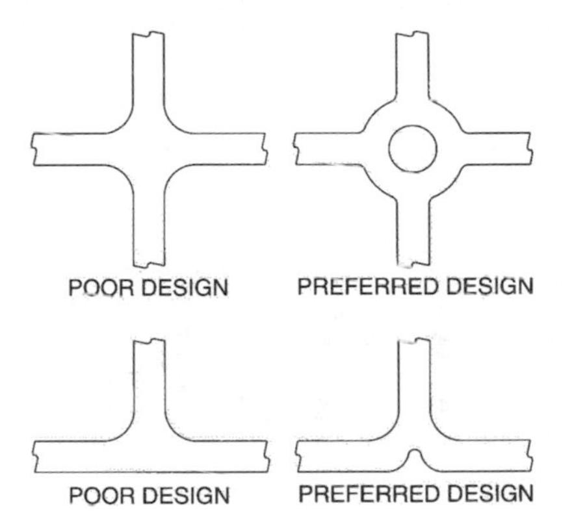

Figure 5-7. Castings must be properly designed to take advantage of their unique characteristics.

can be used to make high volume precision parts such as nuts and is also the starting point for the highest quality tool steels, nickel alloys, refractory metals and cemented carbides. See Figure 5-8.

WELDING FILLER METALS

Welding filler metals (consumables) are melted and become an integral part of the finished joint. They must tolerate very high cooling rates from the molten state and at the same time possess adequate strength and be free from discontinuities. To achieve such properties, welding filler metals generally have different compositions from the base metals

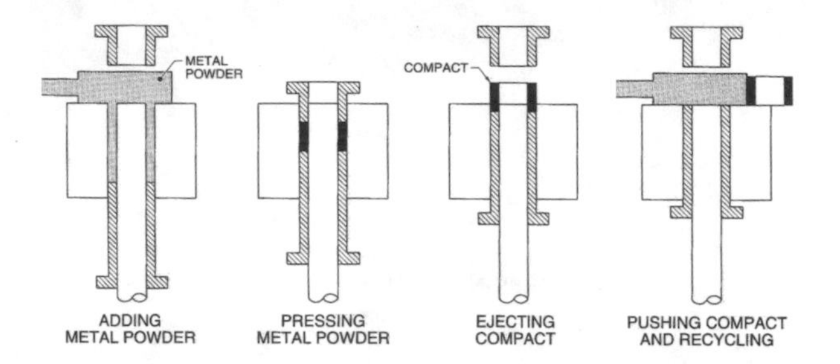

Figure 5-8. Powder metallurgy parts are first compacted and then sintered to densify them.

they join. Welding filler metals have thus evolved as a separate class of materials, just like cast and wrought materials.

Bare electrode, wire, or rod are essentially the same except that rod is wire that is straightened and cut. If the wire is involved in an electrical circuit it is called welding electrode. The deposit chemistry is the composition of the rod or wire itself.

Covered electrode is the most popular type of filler metal used in arc welding. The composition of the covering on the electrode influences the usability of the electrode, the deposit chemistry, and the specification of the electrode. Covered electrodes can be easily damaged and should be treated with care before use, including storage in special heated ovens once the container is broken.

Cored electrode consists of a metal sheath surrounded by a core of fluxing and alloying compounds. Core components perform essentially the same function as the coating in covered electrode. Cored electrode is used for flux-cored arc welding.

Learn the specification system for welding products. See Chapter 3 under AWS standards.

HEAT TREATMENT

Heat treatment is more widely used than cold work to strengthen metals because final properties can be tailored to design requirements and additionally achieve much higher strength levels. Not every metal may be heat treated. For example, the only way of strengthening austenitic stainless steels (300 series) is by cold work.

Heat treatment processes consist of annealing, quenching and tempering, precipitation hardening and case hardening.

Annealing is a general term for various heat treatment processes that reduce internal stress, improve machinability, facilitate cold working, improve corrosion resistance, etc.

Stress relieving (relief) consists of heating a component to a suitable temperature, holding for sufficient time to reduce residual stresses, and cooling slowly enough to minimize development of new residual stresses. Stress relief is commonly used on welded structures, castings, rough machined parts and heat-treated components. See Figure 5-9.

Alloy	ºC	ºF
Gray Cast Iron	538–566	1000–1050
Low Carbon Steel	538–621	1000–1150
Austenitic Stainless Steel	898	1650
Titanium Alloys	482–593	900–1100
Aluminum Alloys	343–399	650–750
Copper Alloys	201–288	375–550

Figure 5-9. Stress relieving temperature ranges vary for different types of alloys. Time at temperature is also a factor.

Stress equalizing is low temperature stress relieving done to balance stresses in cold worked or machined components such as shafts made of stainless steels or nickel alloys. The objective is to avoid movement in service or subsequent fabrication operations.

Spheroidizing consists of heating and cooling to produce a structure that is soft and suitable for machining.

Stabilizing is done on certain corrosion resistant stainless steels and nickel alloys containing titanium, niobium or tantalum to enhance corrosion resistance. They are heated to specific temperatures so that these elements combine preferentially with carbon, preventing any detrimental loss of corrosion resistance. Stabilized alloys include Carpenter® 20Cb-3 and 321 stainless steel.

Solution annealing is done on stainless steels and nickel alloys to develop a soft condition and restore maximum corrosion resistance. It involves heating to a suitable temperature to dissolve all undesirable microconstituents, followed by rapid quenching to prevent re-precipitation of the microconstituents.

Full annealing is done on steels to achieve maximum softness. They are placed in a furnace and taken to a suitable temperature. The furnace is switched off, thus preventing the steel from re-hardening as it cools.

Normalizing is also applied to steels. It consists of heating to a suitable temperature in a furnace, removing, and cooling in still or forced air. Normalizing improves toughness of steel and may be a preparatory step for quenching and tempering.

Quenching and tempering is a heat treatment for steels, tool steels and martensitic stainless steels. It consists of heating to a specific temperature to produce a microstructure of austenite (austenitizing), followed by cooling at a sufficiently rapid rate to make the steel very hard (quenching), and completed by heating to an intermediate temperature to reduce the hardness and recover some ductility and toughness (tempering). Quenching and tempering is the most common heat treatment for alloy and tool steels. For tool steels, double or triple tempering is often required to achieve dimensional stability. See Figure 5-10.

> It is not usually necessary to indicate the heat treatment parameters in the specification, for example, temperature or time at temperature. Final hardness (minimum or maximum) is the most important value to specify. Heat treaters use hardness testing to control their production parameters; how the final hardness is obtained is not usually of concern to the specifier.

Precipitation hardening is a heat treatment that achieves high strength in specific aluminum, copper and nickel alloys, and stainless steels. The alloy is generally supplied in the soft solution annealed condition. After machining, working or stamping to the desired shape it is heated to an intermediate temperature to achieve the desired strength. Various combinations of strength, hardness and toughness are attainable through manipulation of the precipitation hardening temperature. With aluminum alloys precipitation hardening is known as ageing.

> Precipitation hardening alloys should never be used in their solution annealed condition because they lack strength and sometimes ductility, and may be unstable.

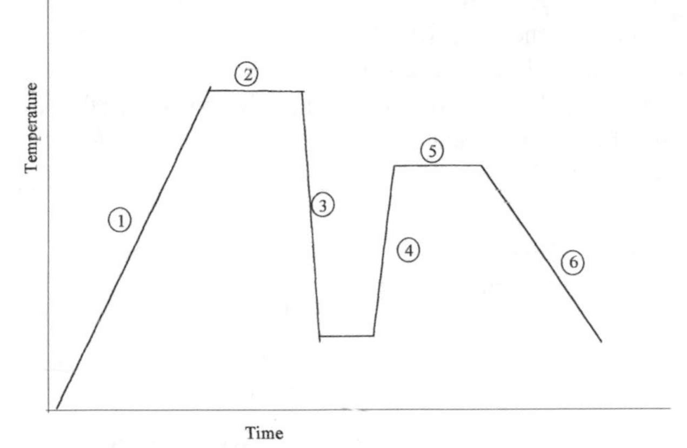

1. Heat up at rate that does not damage part.
2. Hold at austenitizing temperature, e.g. one hour per inch of section.
3. Quench at required rate to achieve desired microstructure, e.g., oil, water.
4. Heat up at any rate.
5. Hold at tempering temperature to obtain required hardness.
6. Cool at any rate.

Figure 5-10. Quenching and tempering is the most common form of heat treatment for most steels.

Case hardening encompasses various heat treatments on steels that develop a thin, hard surface layer on a component, leaving the core (the bulk of the section) relatively strong and tough. Case hardening may provide the optimum combination of properties, for example in shafts, by creating a hard, fatigue resistant surface and a tough, strong body. See Figure 5-11.

Case Hardening Type	Attributes
Carburized	Moderate case depths with excellent capacity for contact loads, good bending fatigue resistance and good seizure resistance
Nitrocarburized and Carbonitrided	Light case depths with fair capacity for contact loads and good resistance to seizure
Nitrided	Light case depths with fair capacity for contact loads, good bending fatigue resistance, excellent seizure resistance and good dimensional control
Induction hardened	Heavy case depths with good capacity for contact loads, good benging fatigue resistance and fair dimensional control
Flame hardened	Heavy case depths with good capacity for contact loads, good bending fatigue resistance and fair dimensional control

Figure 5-11. Case hardening methods are used to increase wear and fatigue resistance of steels.

6
Testing and Identifying Metals

Summary: This chapter describes tests that verify the properties and quality of as received materials. These requirements may be referenced in specifications and performed as checks to prevent materials mix-ups. Types of identification markings used on products are also described.

Many types of nondestructive, leak, corrosion evaluation and mechanical tests may be used to establish and verify quality requirements of the materials specification. In addition, identification markings enable materials to be checked for specification conformance. Field identification methods must be used when all else has failed, but their limitations must be understood.

NONDESTRUCTIVE TESTING

Nondestructive testing or evaluation (NDT or NDE) covers any type of evaluation of a product that leaves it unchanged. NDT techniques are used to qualify that a product is free of specific discontinuities that would impair service performance or lead to failure to meet the purchase specification. Materials specifications indicate the required sensitivity level of specific nondestructive tests for identified quality levels. The object is to avoid over- or under-inspection by selection of the optimum test(s) and test sensitivity.

Discontinuities in products result from inherent material characteristics and from the manufacturing method. For example, casting, as a generic process, is more likely to produce material containing voids (porosity) compared with hot worked product such as plate where the voids are closed up by the mechanical work used to shape the product.

Selection of NDT technique, test parameters and definition of acceptance or rejection criteria are based on the following:

- Sensitivity of NDT technique in detecting specific discontinuities
- Consequence of failure arising from acceptance of the discontinuity
- Criticality of the service

Not all discontinuities in materials are cause for rejection. The controlling specification or standard indicates the size and quantity of permissible discontinuities. When they exceed the threshold level, discontinuities are called defects and are cause for rejection. Do not over inspect because it adds unnecessary cost.

Penetrant Testing (PT) detects surface cracks and porosity and is used on non-absorbent materials (ceramics, glasss, metals and plastics). The surface must be dry and free of contaminants. A penetrant (a colored, light, oil-like liquid), is applied to the surface. It is drawn into any discontinuities open to the surface by capillary action. The penetrant is allowed dwell time to seep in. Penetrant remaining on the surface is removed by wiping or washing, leaving excess penetrant within the cracks or pores. The part is treated with dry powder or a suspension of a powder in a liquid, known as developer. Developer acts like a sponge or blotting paper, drawing penetrant from the discontinuities and enlarging the size of the area of the penetrant indication. See Figure 6-1.

Magnetic Particle Testing (MT) detects surface and subsurface discontinuities such as cracks, but only in ferromagnetic materials (attracted to a magnet). MT is more sensitive than PT in detecting fine cracks such as fatigue, which usually initiate at the surface. The part is magnetized by direct current or rectified alternating current and covered with fine iron powder. It is preferentially attracted to surface or near-to-surface imperfections where the magnetic field leaks out to the surface. There are two methods of applying the iron powder: dry method and the wet method. In the more sensitive wet method, the iron particles are suspended in an oil or water vehicle and made colored or fluorescent in order to maximize contrast. Wet fluorescent magnetic particle testing (WFMP) viewing must be performed under ultraviolet (black) light. See Figure 6-2.

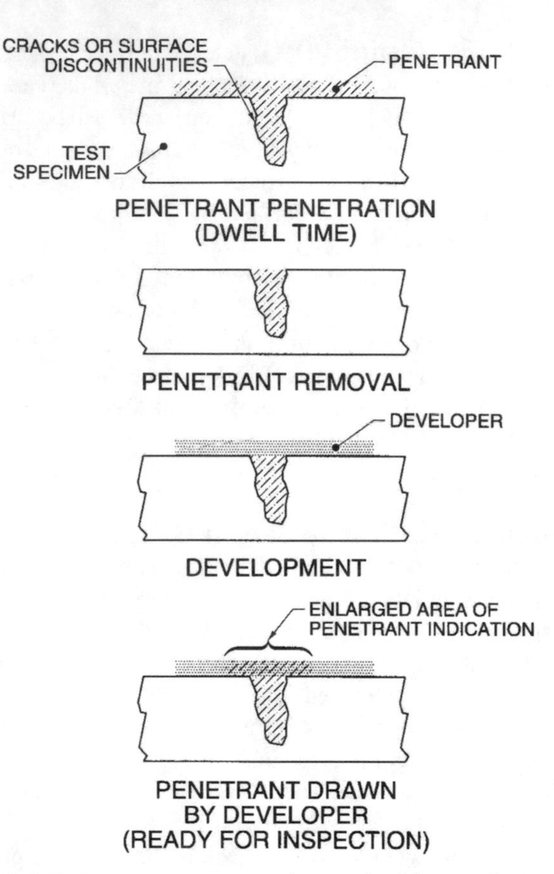

Figure 6-1. Penetrant testing is used primarily to detect surface cracks and porosity.

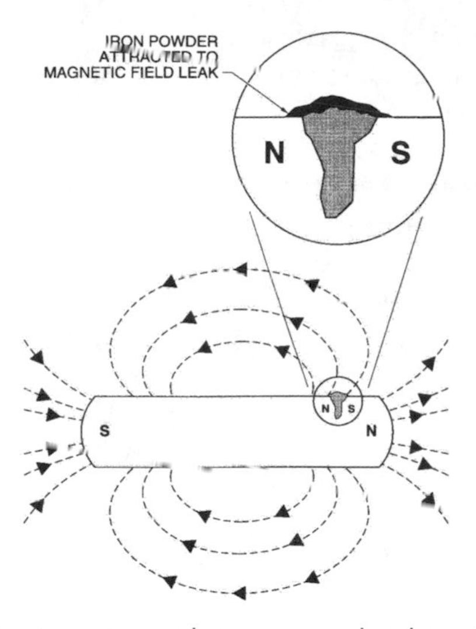

Figure 6-2. Magnetic particle testing is used to detect discontinuities such as fine surface or subsurface cracks.

Ultrasonic Testing (UT) uses electrically produced high-frequency sound waves to penetrate solids and detect discontinuities inside components by means of a transducer containing a piezoelectric crystal: this is a device that converts electrical energy into sound energy and vice versa. The resulting ultrasonic vibrations reflected from within the component are received by the crystal, which converts them into a signal that is amplified, displayed and analyzed. UT is performed on products such as plate, tubing or forgings to ensure they are acceptable and do not contain unacceptable cracks, laminations, seams, inclusions or other discontinuities. See Figure 6-3.

Eddy-current Testing (ET) is an electromagnetic technique for measuring physical and mechanical parameters on

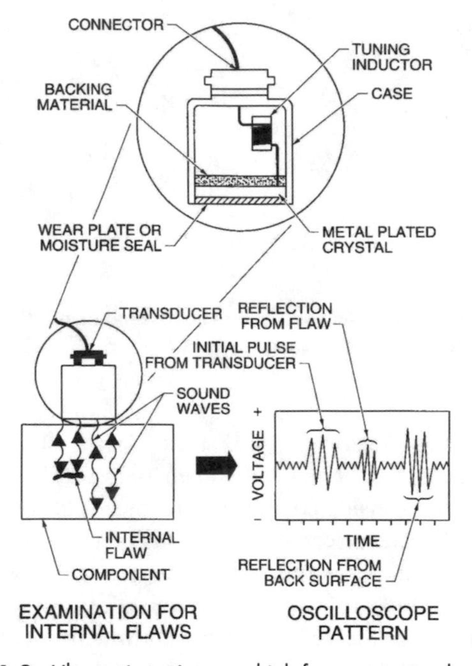

Figure 6-3. Ultrasonic testing uses high frequency sound to penetrate solids and locate discontinuities.

metals, such as surface or near-to-surface imperfections. An alternating current probe is placed near the test specimen and generates fluctuating magnetic fields that induce swirling electric currents within it. The interaction between the probe and the test specimen is displayed as a characteristic pattern on a screen. Analysis of the screen pattern and comparison with standard samples allows ET to detect and measure various discontinuities in products, from fastener composition conformance, to weld discontinuities in pipe and tubing. ET is suited to rapid, mechanized sorting. However, it is susceptible to subtle differences in alloy composition, hardness and heat treated condition that sometimes make it impossible to separate important responses from all the other variables. Comparison with standard samples is critical. See Figure 6-4.

Radiography (RT) utilizes electromagnetic radiation (x-rays and gamma rays) to determine the interior soundness of solids. Radiation from an x-ray generator or a radioactive isotope, such as cobalt-60, passes through the component. It is preferentially absorbed, depending on the density, thickness and contour of the component and to varying degrees by internal discontinuities. The emerging radiation on the opposite side of the component contacts photographic film to produce a radiograph, a shadow picture of the component. Regions of lower density, such as voids, permit more radiation to pass and are seen as darker on the film. RT is primarily used for inspecting welds and castings for internal

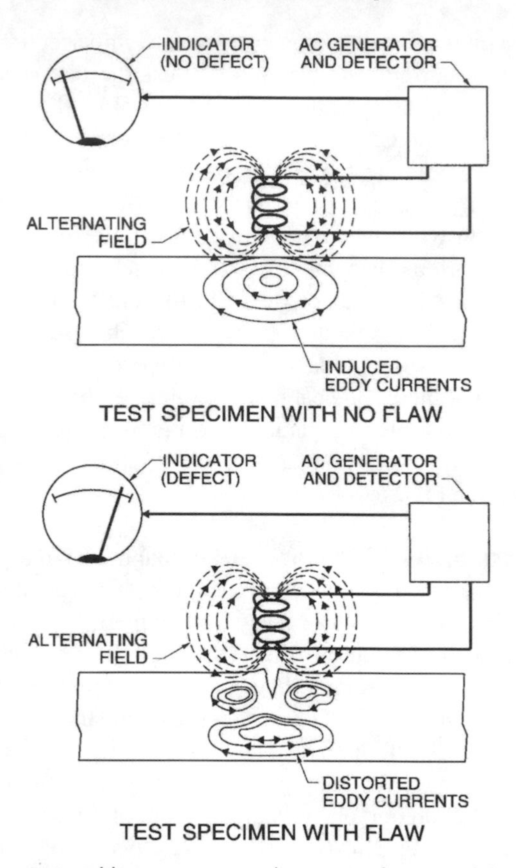

Figure 6-4. Eddy current testing detects specific types of discontinuities in metals but comparison with standards is extremely important.

defects such as porosity, cracks, shrinkage and slag inclusions. See Figure 6-5.

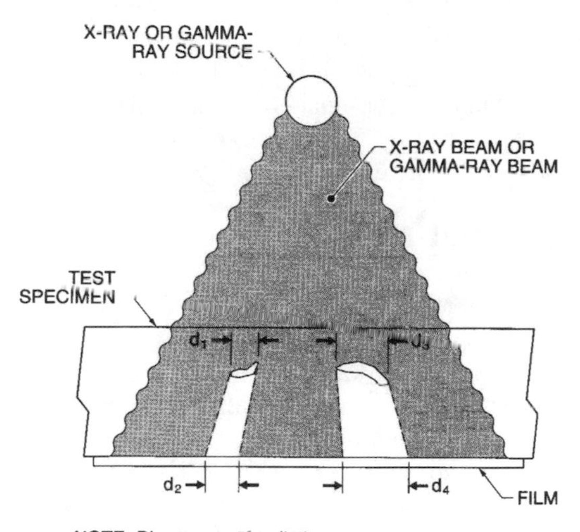

Figure 6-5. Radiography uses electromagnetic radiation to determine the interior soundness of metals.

LEAK TESTING

Leak testing is done to assess the integrity of products or equipment containing a sensitive fluid. Production or manufacturing methods such as welding may create a leak path that is invisible to the naked eye and is revealed by a leak test. Leak testing is included in specifications when necessary. Commonly used fluids for leak testing in order of increasing sensitivity are water, air and helium. The hydrostatic test, in which water is used to fill a structure, is a proof test of the equipment design or of a structural repair. Although a hydrostatic test may reveal leaks it is not designed for that purpose. See Figure 6-6.

MECHANICAL TESTING

Mechanical testing is used to qualify products for conformance with specified properties by means of standard procedures, using standardized, representative product samples. The results are reported on a mill test report (MTR) and indicate whether the product meets mechanical test requirements of the specification test requirements. Mechanical property test reporting is part of the purchase order. Tensile, hardness, and impact are the most common mechanical property tests.

Tensile Test consists of stretching a test specimen in the grips of a test machine. An extensometer, a device for measuring the extension or elongation of the test specimen, is fitted to the test specimen and removed before it breaks. A load-extension (strain) curve is plotted. See Figure 6-7.

Tensile strength is the maximum load applied to the test specimen before it breaks divided by the original cross-sectional

Method	Cost	Complexity	Sensitivity, cc/sec	Comments
Pressure change	Low	Low	10^{-2} to 10^{-4}	Larger systems and smaller leaks require longer times
Chemical	Low	Low	To 10^{-4}	Some methods detect process material (e.g., ammonia)
Bubble detection	Low	Medium	To 10^{-4}	Rapid test for weld seams
Hydrostatic test	Low	Low	10^{-4}	Same sensitivity as bubble method and usually harder to perform
Sonics	Low	Low	10^{-2}	Turbulent flow required for leak detection
Helium mass spectroscopy	High	High	10^{-5} (sniff) 10^{-9} (direct)	Complex test, expensive equipment
Thermal conductivity	Medium	Low	10^{-5}	Highest sensitivity per dollar of equipment cost

Figure 6-6. Leak testing methods have various levels of sensitivity.

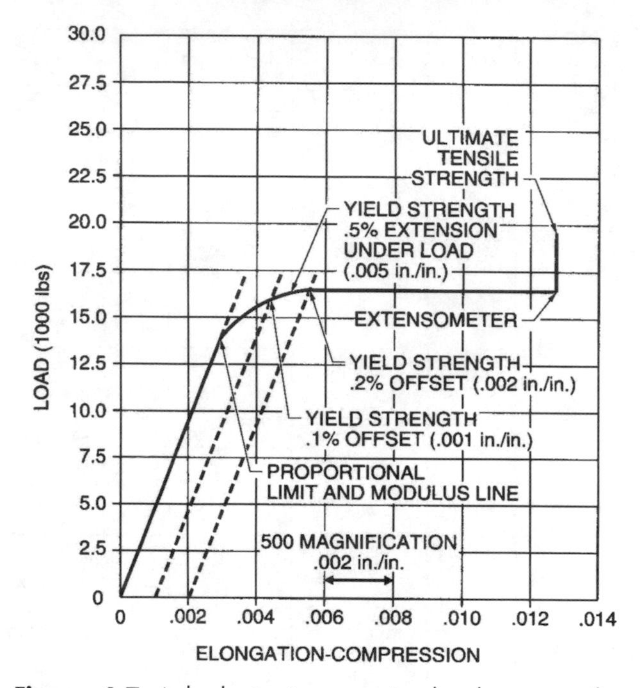

Figure 6-7. A load-extension surve is plotted as a tensile test specimen is slowly pulled.

area of the test specimen. It is measured in thousands of pounds per square inch (ksi) or megapascals (MPa).

Yield strength is the load divided by the original cross-sectional area of the test specimen at a specific point during the tensile test. For carbon steels this point is where a sudden, temporary decrease occurs as the load is applied (yield point). With other metals an artificial yield point is calculated from the load-extension curve by measuring the stress that causes a specific amount of permanent strain, usually .2%. The yield point or yield strength is reported in ksi or MPa.

> Designers incorporate a suitable safety factor with tensile and yield point/yield strength values to develop maximum allowable stress data. For example, in the ASME Boiler and Pressure Vessel Code, maximum allowable stresses are based on one quarter of the tensile strength divided by 3.5, or two thirds of the yield stress/yield point, whichever is lower.

Percent elongation and percent reduction of area are ductility values obtained from the tensile test. They indicate the amount of plastic deformation (stretching) prior to fracture of the test specimen.

Ductility values obtained from the tensile test are not used in design. In service situations the maximum amount of elongation that can usually be tolerated is 1% or 2%. Ductility values are important because they are strongly influenced by the microstructure of the test specimen. Ductility values reveal undesirable microstructure, particularly in heat-treated steels where improper heat treatment could lead to failure or impaired service performance.

Hardness Tests measure resistance of samples or the part surface to indentation or scratching. Indentation hardness tests are the most common. They utilize the surface impression produced by a standardized load to determine hardness. Types of indentation hardness tests include the Brinell, Rockwell and Vickers. There are general precautions to be followed, such a surface preparation and support of the test piece. The Rockwell hardness test is the most used and versatile test. It comes with a variety of attachments, making it useful for measuring the hardness of a wide range of materials in many sizes and shapes. The two most common scales are the Rockwell B (HRB) and the Rockwell C (HRC).

Indentation hardness testing is the most common quality control tool for metals. It is invaluable during heat treatment processes. A metal may be tested as raw stock, re-tested after heat treatment and rough machining, and tested again after finish machining. Hardness values also provide an indication of the amount of cold work a metal has received. For example, the hardness and springiness of alloys such as copper or austenitic (300 series) stainless steels increase with increasing amounts of cold work, but care must be taken in interpreting results. With small amounts of cold work the hardness reading may be misleading because the increase in hardness is not uniform across the section.

> Take great care converting hardness numbers between different scales. ASTM E140 provides conversion tables for different families of alloys. Pocket-sized conversion charts supplied by vendors are usually an extract from the steels portion of ASTM E140. Regular conversion between different hardness readings should be avoided unless there is a large amount of experience and data available to justify making such a decision. See Figure 6-8.

Impact Testing is a qualitative estimate of a material's toughness, which is resistance to catastrophic fracture. Mechanical properties are strongly affected by the rate of loading. A metal tested at a low rate may fracture with a large amount of elongation, but it may fracture with little or no elongation at a high rate. The metal is tough and ductile at the low rate and brittle at the high rate. The test or service temperature and the presence of stress raisers and the thickness of the test specimen also affect toughness. For example, the toughness of certain metals decreases significantly below a characteristic temperature, known as the nil ductility transition temperature (NDTT). The Charpy notched bar impact test is used to provide qualitative information on toughness. It is measured in ft lb energy absorbed to fracture the test specimen at the required test temperature. The higher the energy absorbed the better the toughness. Charpy impact value provides a basis for accepting material and is included in some standards.

> Some specifications require a minimum Charpy v notch impact requirement of 15 ft lb energy absorbed at the minimum expected service temperature. However, this does not mean that a test specimen exhibiting 60 ft lb is four times tougher than the minimum. The main value of notched bar impact testing is as a criterion for acceptance of materials where reliable correlation with service behavior has been obtained.

HARDNESS CONVERSIONS									
		Rockwell Harness Number[b]		Rockwell Superficial Hardness Number					
Brinell Indentation Dia (mm)	Brinell Hardness Number	B-Scale[c]	C-Scale[d]	15-N Scale 15-kg Load	30-N Scale 30-kg Load	45-N Scale 45-kg Load	Shore Scleroscope Hardness	Approximate Tensile Strength (ksi)
2.95	429		45.7	83.4	64.6	49.9	61	217
3.00	415		44.5	82.8	63.5	48.4	59	210
3.05	401		43.1	82.0	62.3	46.9	58	202
3.10	388		41.8	81.4	61.1	45.3	56	195
3.15	375		40.4	80.6	59.9	43.6	54	188
3.20	363		39.1	80.0	58.7	42.0	52	182
3.25	352	*110.0*	37.9	79.3	57.6	40.5	51	176
3.30	341	*109.0*	36.9	78.6	56.4	39.1	50	170
3.35	331	*108.5*	35.5	78.0	55.4	37.8	48	166
3.40	321	*108.0*	34.3	77.3	54.3	36.4	47	160
3.45	311	*107.5*	33.1	76.7	53.3	34.4	46	155
3.5	302	*107.0*	32.1	76.1	52.2	33.8	45	150
3.55	293	*106.0*	30.9	75.5	51.2	32.4	43	145
3.60	285	*105.5*	29.9	75.0	50.3	31.2	-	141
3.65	277	*104.5*	28.8	74.4	49.3	29.9	41	137
3.70	269	*104.0*	27.6	73.7	48.3	28.5	40	133
3.75	262	*103.0*	26.6	73.1	47.3	27.3	39	129
3.80	255	*102.0*	25.4	72.5	46.2	26.0	38	126
3.85	248	*101.0*	24.2	71.7	45.1	24.5	37	122
3.90	241	100.0	22.8	70.9	43.9	22.8	36	118
3.95	235	99.0	21.7	70.3	42.9	21.5	35	115
4.00	229	98.2	20.5	69.7	41.9	20.1	34	111
4.05	223	97.3	*18.8*					
4.10	217	96.4	*17.5*				33	105
4.15	212	95.5	*16.0*					102
4.20	207	94.6	*15.2*				32	100
4.25	201	93.8	*13.8*				31	98
4.30	197	92.8	*12.7*				30	95
4.35	192	91.9	*11.5*				29	93
4.40	187	90.7	*10.0*					90
4.45	183	90.0	*9.0*				28	89
4.50	179	89.0	*8.0*				27	87
4.55	174	87.8	*6.4*					85
4.60	170	86.8	*5.4*				26	83
4.65	167	86.0	*4.4*					81
4.70	163	85.0	*3.3*				25	79
4.80	156	82.9	*0.9*					76
4.90	149	80.8					23	73
5.00	143	78.7					22	71
5.10	137	76.4					21	67
5.20	131	74.0						65
5.30	126	72.0					20	63
5.40	121	69.8					19	60
5.50	116	67.6					18	58
5.60	111	65.7					15	56

[a] 10 mm standard or tungsten carbide ball; 3000 kg load
[b] *italic* entries are beyond normal range; for information only
[c] 1/16" diameter steel ball; 100 kg load
[d] Brale penetrator 150 kg load

Figure 6-8. Interconversion between hardness scales is permitted only if there is sufficient data to make correlations.

CORROSION EVALUATION

Corrosion evaluation is a quality control check to verify that purchased stainless steels and nickel-chromium alloys have been heat treated for optimum resistance, or that their chemical composition will not lead to reduced corrosion resistance in the as-welded state. For example, with stainless steels when it is necessary to determine that an alloy is L grade and thus immune to sensitizing during welding, a corrosion evaluation test is performed. See Chapter 4 under Dual Marked Stainless Steels for explanation of sensitization. Corrosion evaluation testing consists of exposing a sensitized representative sample to a highly corrosive environment under standardized conditions. See Figure 6-9.

Sensitized material that fails corrosion evaluation testing may perform perfectly well in service. The highly oxidizing

Alloy	Test	Typical Pass Criteria
304L	ASTM A262 practice A Or ASTM A262 practice C	No “Ditching” $\ngtr$ 24 mils per year
316L	ASTM A262 practice A Or ASTM A262 practice B	No “Ditching” $\ngtr$ 60 mils per year

Figure 6-9. Corrosion evaluation is a quality control tool that specific alloys are provided to the end user or fabricator in their optimum corrosion resistant condition.

environments used for corrosion evaluation testing may not be anywhere near equivalent to the service environment. Therefore, corrosion evaluation testing should only be performed when it is known that a sensitized alloy would fail prematurely in the service environment.

> Specifying corrosion evaluation tests may lead to significant delay in delivery of material. Suppliers do not normally include such tests as a prerequisite to releasing materials because the majority of the market does not require this form of testing. Now that dual marked stainless steels predominate the market, the requirement for corrosion evaluation should be questioned, except for certain nickel-chromium alloys.

IDENTIFICATION MARKINGS

Identification markings are stamped, embossed or stenciled on metal components and are the first aspect to seek and interpret when examining a metal product.

Foundry Marks are embossed on the exterior of castings such as valve or pump bodies. Foundry marks are incorporated into the pattern when the mold is made and are easily visible. The ASTM grade number, foundry name or logo, heat number, and foundry shorthand for the alloy type are generally included. Additional information such as design pressure and temperature rating of the casting may also be included. Query the manufacturer for his internal designation system. See Figure 6-10.

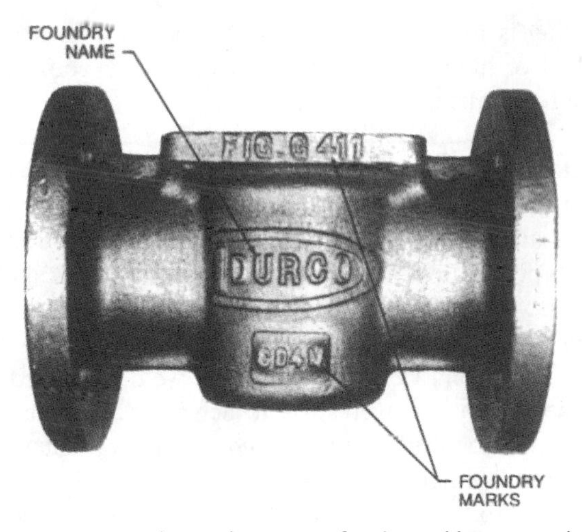

Figure 6-10. Foundry marks consist of embossed letters, numbers, or logos that manufacturers use to identify cast products.

Color Coding is a method of identification that most often uses colored striping painted at one end of the product before temporary storage. Color coding schemes allow simple identification of metals stored at a specific location to make retrieval foolproof. For this reason there is no universal color coding system. Whenever a piece is cut from the product, the color coding must be restored to the cut end.

Certain chemical elements in some paints and marking materials used for color coding are potentially harmful. They may cause catastrophic cracking or pitting on susceptible alloys, particularly stainless steels and nickel alloys.

Cracking is more likely to occur on exposure to heat (eg, welding or high-temperature service) or to specific corrosive environments. See Figure 6-11.

> Chlorine, sulfur and zinc are potentially harmful chemical elements. Paints and crayons used to mark susceptible alloys must contain low quantities (measured in parts per million, or ppm) of the harmful chemical elements. Less than 100 ppm is allowable. Even if approved marking materials are used, they should be removed from areas that are to be welded, brazed or soldered. An approved solvent such as a non-chlorinated type should be used to remove marking materials. The same principles apply when adhesive-backed tapes are used to fix items to stainless and nickel alloy products (e.g., for radiography). All traces of adhesive must be removed from the surface with an approved solvent. There are no restrictions on using fiber tip markers because they do not leave solid residue. Solvent removal is not required with fiber tip markers.

Stencil Marking is printed marking on wrought products that indicate alloy type, standard designation (e.g., ASTM)), dimensions and heat number. Stencil markings are repeated at regular intervals across the product so identification is not lost when it is sectioned. Stencil markings are not permanent and may degrade with outdoor exposure or during service.

Figure 6-11. Stencil markings are used on wrought products and are usually applied during the manufacturing process.

Stamping/Embossing are impressions on items like forgings or fasteners made by a metal die. With forgings the identification consists of the standard (ASTM), material type, the pressure and temperature rating of the forging, and the forge shop logo. The Drop Forgings Association maintains a list of logos. Fasteners are identified by an embossed or stamped marking on the head, or at one end in the case of studs. Space is restricted on fasteners so that a code is used to identify the ASTM type or other material type and strength level. The manufacturer's logo is also included. The Industrial Fasteners Institute publishes a list of manufacturers' logos. See Figure 6-12.

> Stamped identifications on structural components must be located so as to avoid introduction of surface mechanical stresses that lead to cracking during service. Locations such as fillets on shafts are unacceptable. As a further precaution, low-stress stamping should be used. Low-stress stamping dies have broken (rounded) corners on the impression symbols to minimize the intensity of the localized stress caused by the stamping impression.

Welding Filler Metals (Wire and Electrodes) use various marking systems. For covered electrodes (welding filler metal with a flux coating) the AWS designation is stenciled on the flux coating at one end of the electrode. For bare wire the AWS designation is printed on a paper flag that is glued to one or both ends of the wire. AWS publishes *Filler*

Figure 6-12. Fasteners and forgings are stamped or embossed with identification markings.

Metals Comparison Charts that correlates manufacturers trade names with standard AWS designations. See Figure 6-13.

FIELD IDENTIFICATION OF METALS

Field identification provides rapid, nondestructive verification of the composition of metal products before they are fabricated into components, placed in service, or stored for later use. Field identification may be required when accompanying paperwork is lost or identification markings are obliterated or not to be trusted.

> Many metals look alike. It is relatively easy to substitute the wrong metal when the identification has been obliterated or the accompanying paperwork lost. For example, chrome-moly steels look like carbon steel, are magnetic, and rust like carbon steel if stored outdoors. Chrome-moly steels are used in critical applications such as piping for handling steam or hydrogen. Carbon steel should not be mistakenly substituted for chrome-moly steel in a critical application, because it might fail catastrophically.

Color and Density. When faced with an unknown metal having no markings, the first features to evaluate are its color, and, if it can be picked up, its density. Although by no means exact, they can provide initial information, especially if known samples are available for comparisons. See Figure 6-14.

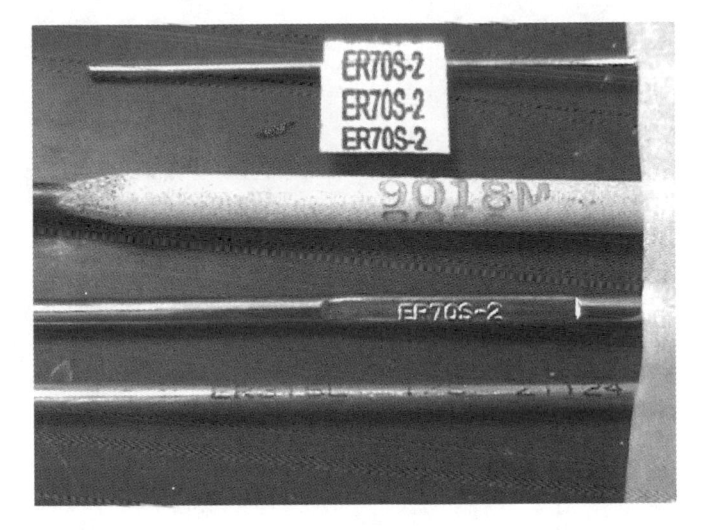

Figure 6-13. Welding filler metals are identified by a stenciled AWS designation at the end of a coated electrode and a paper flag at the end of bare wire.

Color	Metal
Red or reddish	Copper, >85% copper alloys
Light brown or tan	90-10 copper-nickel
Dark yellow	Bronzes and gold
Light yellow	Brasses
Bluish or dark gray	Lead, zinc, and zinc alloys
Silvery white with soft luster	Aluminum
Silvery white with bright luster	Stainless steels
Gray	Carbon and low-alloy steels, 70-30 copper-nickel
White or light gray	Nearly all others

Density Grouping	Density Range (g/cc)	Metals in Range	
Very high	12 to 22	Gold Iridium Osmium Palladium Platinum	Rhodium Ruthenium Tantalum Tungsten Uranium (depleted)
High	9.8 to 11.9	Bismuth Lead	Molybdenum Silver
Average	6 to 9.7	Antimony Cadmium Cast irons Copper alloys Nickel	Nickel alloys Stainless steels Steels Tin Zinc
Low	1 to 5.9	Aluminum alloys Beryllium	Magnesium alloys Titanium alloys

Figure 6-14. Color and density provide relative comparisons of unknown metals.

Magnetism. Metals are broadly sorted by applying a magnet to the surface and testing for an attractive force. See Figure 6-15. There are a few caveats, however

- The magnetism of a metal can change with temperature. For example, the Curie temperature of Monel 400 is 20–50°F. Above this temperature, it will exhibit little or no magnetism.

Magnetic Behavior Grouping	Applicable Metals
Strong Attraction	Steels: carbon, alloy, tool Cast Irons: gray, ductile, malleable Cobalt Nickel Stainless Steels: ferritic, duplex, martensitic, martensitic precipitation hardening
Weak Attraction	Stainless Steels: cold worked 302 and 304, 308 filler metal Monel (becomes nonmagnetic in boiling water)
No Attraction	Commercially pure nonferrous metals (except nickel and cobalt) Copper-nickels Hastelloys, Inconels, Incoloys Stainless steels: austenitic, austenitic precipitation hardening Stellite

Figure 6-15. Metals can be grouped by their response to a magnet.

- Magnetic properties of some metals may change with their mechanical history. For example, 302 and 304 stainless steels are nonmagnetic in the annealed (soft) condition, but become increasingly magnetic as they are cold worked.
- Misidentification of stainless steel castings and weld filler metals is the most common error with magnetic sorting. The 300 series stainless steels (wrought products) are nonmagnetic, with the exceptions noted above. However, their cast counterparts (CF-3M for 316L and CF-3 for 304 stainless steel) are slightly magnetic. This is because slight compositional changes are made to castings to improve their castability. The compositional changes result in the slight magnetism that is observed in cast products. Stainless steel weld filler metals are also slightly magnetic for the same reason.

In spite of such complications, magnetic sorting is used constantly in the field. Magnets can also be useful identification tools in conjunction with sophisticated quantitative instruments. For example, x-ray fluorescence analysis of a component showed it to be 100% chromium. This is because the technique utilizes only a very thin surface layer. A test with a magnet indicated attraction. It was concluded that the part was chrome plated and not solid chromium (which would have been very unlikely).

Chemical Spot Testing identifies metals by the color changes that occur when the surface is contacted with a specific chemical. The specific chemical is applied either directly to the metal, or to a solution produced by dissolving a small amount of the metal on a filter paper soaked with a specific chemical. The solution is usually produced with the assistance of an electric current, in which case it is known as the electrographic technique. See Figure 6-16.

Chemical spot tests utilize a decision tree to identify the alloy through a series of chemical applications. Many of them are straightforward, and with some experience can be quite reliable, but other tests fall into ambiguous "gray" zones, generating questionable results.

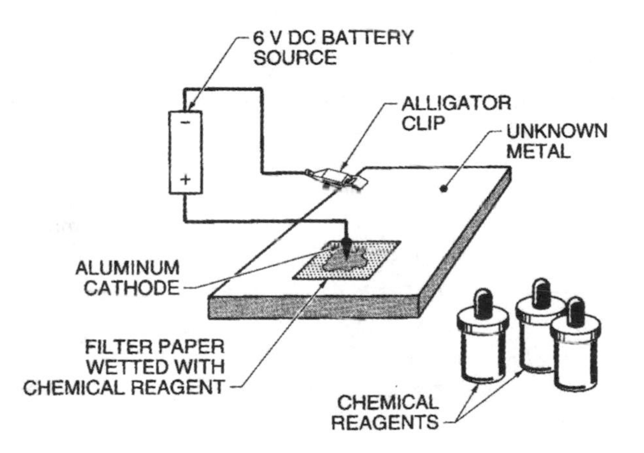

Figure 6-16. The electrographic test is the commonest chemical spot test.

For some applications chemical spot testing may be preferred over more sophisticated methods. An example where chemical spot testing proved useful was in a chemical process that was originally designed and built completely out of Monel 400. Over the years, much of the piping and process equipment was replaced with stainless steel and all the equipment had been painted. It became necessary to identify which system components were stainless steel and which were Monel 400. It would have taken roughly three weeks to complete the identification with an x-ray fluorescence analyzer, but with the use of a simple chemical spot test the job was completed in significantly less time.

Spark Testing identifies carbon steels, low alloy steels, and tool steels by visual examination of the spark stream when the metal is held against a grinding wheel rotating at high speed. The chemical composition of the metal influences the form of spark stream produced. Variations in carbon content or other specific alloying elements may be detected by the changes in the spark stream. Small portable grinders are used because they can be transported to the field. The wheel must rotate at high speeds (between 7500 fpm to 15,000 fpm) and be hard (940 grain alumina). Before testing, the grinding wheel is dressed with a diamond wheel dresser to remove particles of metal from the previous test. Steady pressure is applied to produce a uniform stream about 1 foot horizontally to the line of vision. See Figure 6-17.

Metal	Stream Volume	Relative Length[a]	Color of Stream	Color of Streaks	Quantity of Spurts	Nature of Spurts
1. Wrought iron	Large	65	Straw	White	Very few	Forked
2. Machine steel (AISI 1020)	Large	70	White	White	Few	Forked
3. Carbon tool steel	Moderately large	55	White	White	Very many	Fine, repeating
4. Gray cast iron	Small	25	Red	Straw	Many	Fine, repeating
5. White cast iron	Very small	20	Red	Straw	Few	Fine, repeating
6. Annealed mall. iron	Moderate	30	Red	Straw	Many	Fine, repeating
7. High-speed steel (18-4-1)	Small	60	Red	Straw	Extremely few	Forked
8. Austenitic manganese steel	Moderately large	45	White	White	Many	Fine, repeating
9. Stainless steel (Type 410)	Moderate	50	Straw	White	Moderate	Forked
10. Tungsten-chromium die steel	Small	35	Red	Straw[b]	Many	Fine, repeating[b]
11. Nitrided Nitralloy	Large (curved)	55	White	White	Moderate	Forked
12. Stellite	Very small	10	Orange	Orange	None	
13. Cemented tungsten carbide	Extremely small	2	Light orange	Light orange	None	
14. Nickel	Very small[c]	10	Orange	Orange	None	
15. Copper, brass, and aluminum	None				None	

[a] actual length varies with grinding wheel, pressure, etc. [b] blue-white spurts [c] some wavy streaks

Norton Company

Figure 6-17. Spark charts are compared with spark stream characteristics to identify unknown metals when using the spark test.

Optical Emission Spectroscopy consists of sparking of a metal surface by means of an oscillating electric current. Light composed of various wavelengths is emitted from the surface. Chemical elements in the metal determine the component wavelengths produced, and the intensity of each component wavelength is proportional to the concentration of its corresponding chemical element. The light is collected through a prism and viewed as a series of colored lines. These are compared with standards for different alloys to identify the unknown metal. OES is portable and allows rapid sorting by an experienced operator. See Figure 6-18.

X-Ray Fluorescence Spectroscopy (XRF) provides rapid, quantitative analysis of specific chemical elements in

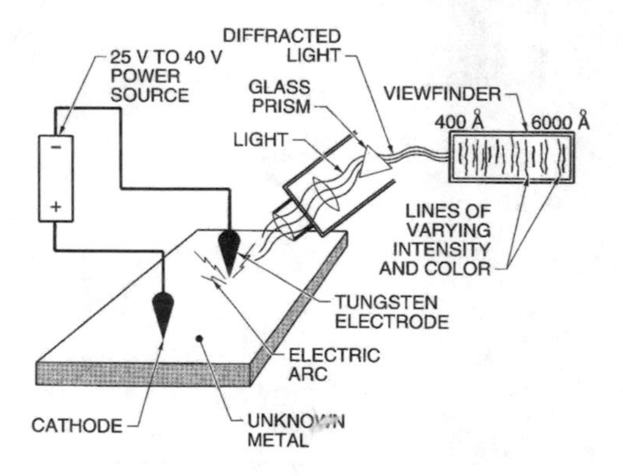

Figure 6-18. Optical emission spectroscopy uses the light emitted from unknown metal surfaces to identify the metal.

an unknown alloy. It may also be used to provide qualitative verification by sorting correct materials from improper ones. The surface is exposed to low level radiation and the alloy, in responding to the radiation, emits fluorescent x-rays that have energy levels and wavelengths characteristic of the specific chemical elements in the unknown metal. They pass through a detector and within 1–5 minutes provide quantitative analysis. The instrument is limited to a specific number of elements, for example 21 elements with one type of model. Carbon, unfortunately, cannot be detected by any type of XRF instrument. Therefore, none can distinguish between various carbon steels or low and regular carbon stainless steels. However, the instrument is a powerful tool and has more than paid for itself by detecting significant metal mix-ups that could have led to serious incidents, for example during construction of a new plant. Another advantage of the technique is that it does not generate sparks and may be used without explosion permit. See Figure 6-19.

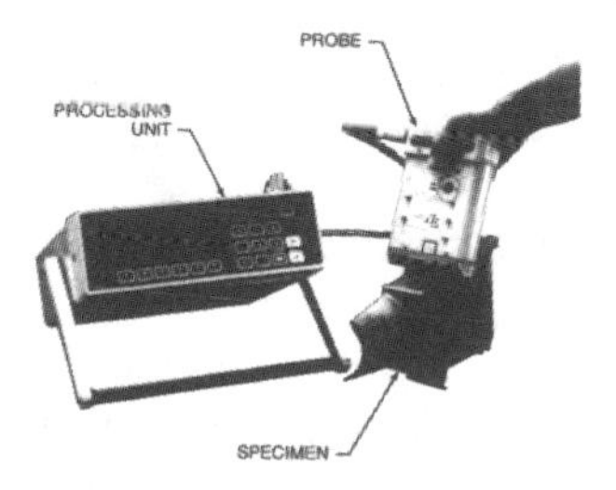

Figure 6-19. With x-ray fluorescence spectroscopy elemental analyses may be obtained for unknown metals.

7
Non-Metals

Summary: This chapter describes how plastics, rubbers and cemented carbides are categorized, specified and identified. Specifying by part number from approved suppliers is the most reliable way of preventing materials mix-ups with non-metals.

Non-metals are used for corrosion and wear resistance in piping, tanks, valve seats, gaskets, seals, etc. They are generally procured by trade names, although generic specifications exist for commodities such as PVC or plastic-lined pipe. Non-metals include plastics, rubbers, and cemented carbides.

PLASTICS AND RUBBERS

Plastics are based on repeating units of a monomer, such as ethylene, to create a polymer (e.g., polyethylene). The principal limitations of plastics compared with metals are their relatively low strength at elevated temperatures and limited ductility. Many plastics have outstanding chemical resistance.

Plastics consist of high molecular weight polymer chains that are generally synthesized from low molecular weight monomers. If the polymer chains are not adequately cross-linked they may soften and creep by heating. Plastics are thus classified into two groups: thermoplastics (few crosslinks) and thremosets (highly crosslinked). Rubbers are a class of plastics mostly composed of thermosets.

Thermoplastics have low ductility, but can be made more pliable with the addition of plasticizers. Once formed, thermoplastics may be repeatedly reshaped with heat and pressure. See Figure 7-1.

Thermosets usually contain fillers of various kinds to modify their properties. Once formed and cured, they cannot be reshaped, re-melted or reprocessed. See Figure 7-2.

Rubbers (elastomers) are generally thermoset materials that exhibit high elongation to break and good recovery. These properties make them suitable for gaskets, seals, expansion joints, linings, diaphragms etc. One or more monomers comprise the polymer backbone. They may polymerize in a random or alternating fashion. Fillers, curing agents, and additives are incorporated to make the final product. Carbon black is the most common filler. Rubbers are most often identified by the trade name of the monomer upon which they are based and their as-cured hardness, e.g.,

(text continued on page 159)

Thermoplastic Type	Designation	Max. Continuous Service Temperature	Comments
Acrylonitrile-butadiene -styrene	ABS	90°C	Low price and limited chemical resistance. Useful for gas, water and waste piping
High density polyethylene	HDPE	60°C	Most widely used for drums, liners, piping and components
Polypropylene	PP	65°C	Stronger and more chemically resistant than HDPE with similar uses
Polyvinyl chloride	PVC	65°C	For piping and as a lining
Chlorinated polyvinyl chloride	CPVC	80°C	Similar to PVC but with better temperature rating
Poly tetra fluoro ethylene	PTFE	260°C	Almost immune to chemical attack. A lining for piping and valves, valve seats, flexible hose, gaskets. Consists of granules bonded together which do not fuse on heating.
Perfluoro alkoxy oxane	PFA	260°C	Similar to PTFE, but can be melted, molded and welded

Perfluoro ethylene propylene	FEP	175°	Like PTFE but can be melted and welded
Ethylene tetrafluoro ethylene	ETFE	150°F	Less chemical resistance than PTFE but fully formable into many products
Poly vinylidene fluoride	PVDF	175°C	Less corrosion resistance than PTFE but cheaper and weldable
Polyether ether ketone	PEEK	315°C	For high temperature parts and wear resistance, e.g., bushings and valve seats
Polypherelyne sulfide	PPS	315°C	Similar to PEEK
Polyimide	PI	315°C	Similar to PEEK
Nylon	PA	175°C	Very limited chemical resistance, for mechanical parts

Figure 7-1. Thermoplastics may be repeatedly reshaped with the application of heat and pressure.

Thermoset Type	Designation	Max. Continuous Service Temperature	Comments
Epoxy	EP	100°C	Usually reinforced with fiberglass and used for commodity piping. Also used as a lining
Phenolic	PF	175°C	Reinforced with graphite or glass fiber and used for structural parts. Also used as a lining and impregnation for graphite equipment
Furan	FF	175°C	Best chemical resistance of thermosets, can be reinforced with fiberglass for structural applications, but difficult to mold
Polyester		95°C	Easy to mold with fiberglass and used for many types of FRP equipment and piping
Vinyl ester	VE	95°C	Superior to reinforced polyester

Figure 7-2. Once formed and cured, thermosets cannot be reshaped, re-melted, or reprocessed.

(text continued from page 155)

EPDM (monomer name), 75 Shore A (as cured hardness). However, this type of identification scheme is insufficient to guarantee reliable service performance in critical applications. See Figure 7-3.

SPECIFYING PLASTICS AND RUBBERS

Plastic and rubber parts should be specified by the manufacturer's designation and not by generic or standardized descriptors. For example, the registered trademark Viton®, developed by DuPont, is inappropriately represented by rubber suppliers when they incorporate only a small amount of genuine Viton® in their product. The result is there are many products with the given name Viton, yet none of which have adequate or equivalent properties to a product based on 100% virgin Viton®. DuPont Dow Elastomers produces over 90 specific types of genuine Viton and controls the manufacturing process to create products with predictable properties (e.g., Viton A, Viton GF).

Specifying rubber parts by generic product description invariably results in material of unknown pedigree that may fail prematurely.

The automotive and aerospace industries employ rigid classification systems to describe all the ingredients in the product and key performance requirements. They are defined in ASTM D2000 for rubbers and D4000 for plastics.

Rubber Type	Designation	Max. Continuous Service Temperature	Comments
Natural rubber	IR	66°C	Linings, hose, gaskets
Neoprene	CR	82°C	Linings, hose, gaskets, o-rings
Ethylene propylene diene	EP	177°C	Lining, hose, gaskets, parts, o-rings
Chlorobutyl	CHR	120°C	Linings
Silicone	FVSI	200°C	Minimal chemical resistance
Fluorosilicone	FVMQ	175°C	Good chemical resistance
Polysulfide	T	105°C	For expansion joints in concrete
Chlorosulfonated polyethylene	CSPE	93°C	Pond linings
Fluoroelastomer	FKM	300°C	O-rings, gaskets, seals, parts
Fluoroelastomer	FFKM	300°C	O-rings, gaskets, seals parts

Figure 7-3. Rubbers are materials with high elongation to break and good recovery.

The process industry has yet to adopt a rigid methodology for classifying plastics and rubbers. Consequently, it is imperative for the user to specify critical parts by manufacturer's designation number and purchase them only from approved suppliers. See Figure 7-4.

IDENTIFYING PLASTICS AND RUBBERS

Unlike metals, it may be difficult or impossible to obtain a complete analysis of a plastic or rubber compound. Elemental analysis of plastics and rubbers will not reveal how the elements were arranged (their architecture), which is the key to their performance. However, it may be possible to identify the building blocks (monomers) that constitute the architecture of plastics and rubbers and compare them with standard, known products for a reasonably accurate identification.

Thermal analysis provides information that may be used to characterize plastics and rubbers. Identification tests that involve combustion of a test sample are also used, but may produce hazardous vapors.

Thermal Analysis involves differential scanning calorimetry (DSC) to identify various common engineering plastics. A thumbnail size sliver of the unknown material must be obtained from a location that will not impact the performance of the part, for example the outside edge of the gasket surface on a lined pipe or valve. The sample is sent to a laboratory that can perform DSC. See Figure 7-5.

Checklist	✓
ASTM Identification (e.g., FKM)	
Rubber Family name (e.g., Viton® A)	
Manufacturer	
Manufacturer's Compound code	
Key Mechanical Property (e.g., hardness Shore A)	
Cure or Other Descriptor (e.g., peroxide cured)	
Certification Required (yes or no)	
Other Special Requirements	

Figure 7-4. Specification format for plastic and rubber parts.

SUPPLY CHAIN FOR RUBBER PARTS

Rubber parts are extremely susceptible to materials mix-ups. The complexity of the supply chain between the rubber pro-ducer and the part user is the principal reason. Carbon black is the most effective filler material for optimum mechanical and chemical resistant properties. Consequently, most rubber parts are black, which is a serious impediment to positive materials identification. The end user must man-

Plastics which can be identified using DSC:
PVC, polyvinyl chloride
HDPE, high density polyethylene
PP, polypropylene
PVDF, polyvinylidene flouride, Kynar®
CTFE, chloro tetrafluoro ethylene, Kel-F®
ETFE, ethylene tetra fluoro ethylene, Halar®
PFA, perfluoro alkoxy oxane, Teflon® PFA
FEP, Perfluoro ethylene propylene, Teflon® FEP
PTFE, poly tetra fluoro ethylene, Teflon® TFE
ETFE, ethylene tetrafluoro ethylene, Tefzel®

Use the following procedure to make an identification. Remove a sliver of the unknown plastic about the size of a thumbnail clipping from a location that will not affect the performances of the part. For example, from the outside edge or a gasket surface or on a lined pipe or valve. Send the sample to a qualified laboratory to carry out the identification.

Figure 7-5. Differential scanning calorimetry (DSC) can be used to identify many common engineering plastics in the laboratory.

age the supply chain by developing specifications that indicate the specific rubber compounds, the manufacturer's part numbers and by ordering products only from approved suppliers. Approved suppliers must verify that their products comply fully with the purchase specification and are not compromised in any way. See Figure 7-6.

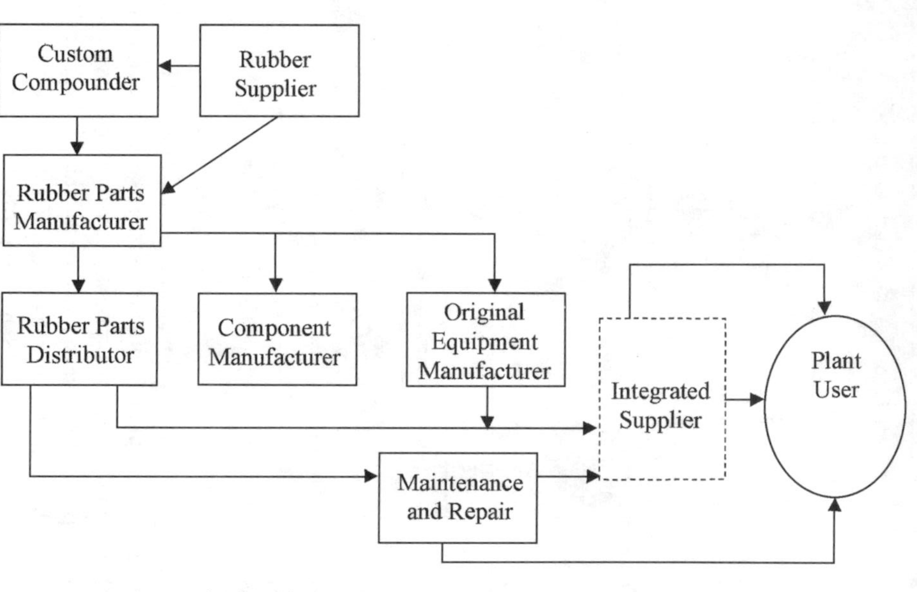

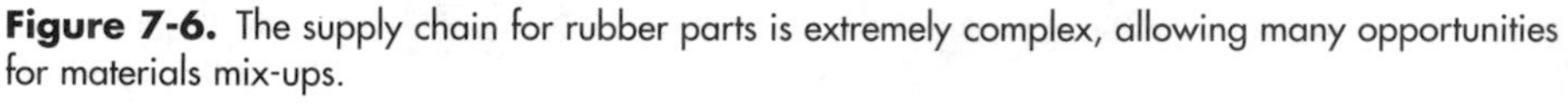

Figure 7-6. The supply chain for rubber parts is extremely complex, allowing many opportunities for materials mix-ups.

Roles and responsibilities in the supply chain of rubber parts are as follows:

- *Rubber Supplier.* Manufactures rubber gums and sells them to fabricators
- *Custom Compounder.* Mixes rubber gums with other compounding ingredients to create compounds. The compounds may be proprietary to the compounder or follow a customer's specific formulation.
- *Rubber Parts Manufacturer.* May mix his own proprietary compound or purchase one from a custom compounder. The manufacturer then fabricates finished rubber products.
- *Rubber Parts Distributor.* Distributes a broad range of rubber products, usually within a specified geographical location. Represents one or more rubber parts manufacturers of a part (e.g., o-ring, gasket, seal, etc.).
- *Component Manufacturer.* Designs and assembles systems that may include rubber parts (e.g., mechanical seals, expansion joints, valves).
- *Original Equipment Manufacturer (OEM).* Manufactures and sells finished products such as pumps, valves, compressors, vessels, etc., for which OEM's also supply spare parts.
- *Maintenance and Repair.* Provides supply and service functions directed to immediate plant or operational needs
- *Integrated Supplier.* Service companies contracted to purchase, quality check, inventory and supply plants with many types of parts, from fasteners to pipe fittings and gaskets.

CEMENTED CARBIDES

Cemented carbide parts are used for demanding wear applications. They are primarily made of tungsten carbide and contain a small percentage of binder such as nickel or cobalt to provide structural homogeneity. Cemented carbides products include knives, blades, nozzles, jets guides, valve seats, seal parts, etc. The properties and design of cemented carbides strongly influence part life and overall equipment reliability.

SPECIFYING CEMENTED CARBIDES

Material inconsistencies are common with cemented carbides. They most often occur when purchasing supposedly identical components from different suppliers and specification details have not been indicated. Unlike ASM or UNS designations that exist for metals, there is no consistent industry standard for cemented carbides, despite their industrial importance. The industry classifies cemented carbides into very general use categories. Consequently, changing from one commercially available product to another may result in very different performance by the part. To obtain consistent performance one must go beyond the traditional industry general use categories and specify features such as grain size, binder type and content, hardness, method of compaction, etc. See Figure 7-7.

Requirement	Units or applicable standard
Commercial trade name	Per manufacturer or supplier
Hardness	Rockwell A (HRA) per ASTM B294
Tensile rupture stress	Ksi, min., per ASTM B406
Density	G per cc, per ASTM B311
Binder type	e.g., nickel, cobalt
Binder content	%
Grain size	e.g., coarse, medium, fine, submicron, ultra fine submicron
Porosity	% per ASTM B276
Submicron amount	%
Eta phase	None visible
Compaction method	e.g., sinter HIP, or HIP at 14 ksi min
Finish of cutting surface	rms
Percentage of recycle carbide	None used

Figure 7-7. Proper specification of carbides helps ensure that no deviations occur and part perform predictably.

Appendix

List of Chemicals

OSHA 1910.119 lists highly hazardous chemicals and permissable quantities for which for regulations apply.

OSHA Regulations (Standards - 29 CFR)

List of Highly Hazardous Chemicals, Toxics and Reactives (Mandatory). - 1910.119 App A

◀ OSHA Regulations (Standards - 29 CFR) - Table of Contents

- **Standard Number:** 1910.119 App A
- **Standard Title:** List of Highly Hazardous Chemicals, Toxics and Reactives (Mandatory).
- **SubPart Number:** H
- **SubPart Title:** Hazardous Materials

This Appendix contains a listing of toxic and reactive highly hazardous chemicals which present a potential for a catastrophic event at or above the threshold quantity.

CHEMICAL NAME	CAS*	TQ**
Acetaldehyde	75-07-0	2500
Acrolein (2-Popenal)	107-02-8	150
Acrylyl Chlorde	814-68-6	250
Allyl Chlorid	107-05-1	1000
Allylamine	107-11-9	1000
Alkylaluminum	Varies	5000
Ammonia, Anhydrous	7664-41-7	10000
Ammonia solutions (greater than 44% ammonia by weight)	7664-41-7	15000
Ammonium Perchlorate	7790-98-9	7500
Ammonium Permanganate	7787-36-2	7500
Arsine (also called Arsenic Hydride)	7784-42-1	100
Bis(Chloromethyl) Ether	542-88-1	100
Boron Trichloride	10294-34-5	2500
Boron Trifluoride	7637-07-2	250
Bromine	7726-95-6	1500
Bromine Chloride	13863-41-7	1500
Bromine Pentafluoride	7789-30-2	2500
Bromine Trifluoride	7787-71-5	15000
3-Bromopropyne (also called Propargyl Bromide)	106-96-7	100
Butyl Hydroperoxide (Tertiary)	75-91-2	5000
Butyl Perbenzoate (Tertiary)	614-45-9	7500
Carbonyl Chloride (see Phosgene)	75-44-5	100
Carbonyl Fluoride	353-50-4	2500

Cellulose Nitrate (concentration greater than 12.6% nitrogen	9004-70-0	2500
Chlorine	7782-50-5	1500
Chlorine Dioxide	10049-04-4	1000
Chlorine Pentrafluoride	13637-63-3	1000
Chlorine Trifluoride	7790-91-2	1000
Chlorodiethylaluminum (also called Diethylaluminum Chloride)	96-10-6	5000
1-Chloro-2,4-Dinitrobenzene	97-00-7	5000
Chloromethyl Methyl Ether	107-30-2	500
Chloropicrin	76-06-2	500
Chloropicrin and Methyl Bromide mixture	None	1500
Chloropicrin and Methyl Chloride mixture	None	1500
Commune Hydroperoxide	80-15-9	5000
Cyanogen	460-19-5	2500
Cyanogen Chloride	506-77-4	500
Cyanuric Fluoride	675-14-9	100
Diacetyl Peroxide (concentration greater than 70%)	110-22-5	5000
Diazomethane	334-88-3	500
Dibenzoyl Peroxide	94-36-0	7500
Diborane	19287-45-7	100
Dibutyl Peroxide (Tertiary)	110-05-4	5000
Dichloro Acetylene	7572-29-4	250
Dichlorosilane	4109-96-0	2500
Diethylzinc	557-20-0	10000
Diisopropyl Peroxydicarbonate	105-64-6	7500
Dilauroyl Peroxide	105-74-8	7500
Dimethyldichlorosilane	75-78-5	1000
Dimethylhydrazine, 1,1-	57-14-7	1000
Dimethylamine, Anhydrous	124-40-3	2500
2,4-Dinitroaniline	97-02-9	5000
Ethyl Methyl Ketone Peroxide (also Methyl Ethyl Ketone Peroxide; concentration greater than 60%)	1338-23-4	5000
Ethyl Nitrite	109-95-5	5000
Ethylamine	75-04-7	7500
Ethylene Fluorohydrin	371-62-0	100
Ethylene Oxide	75-21-8	5000
Ethyleneimine	151-56-4	1000
Fluorine	7782-41-4	1000
Formaldehyde (Formalin)	50-00-0	1000
Furan	110-00-9	500
Hexafluoroacetone	684-16-2	5000
Hydrochloric Acid, Anhydrous	7647-01-0	5000
Hydrofluoric Acid, Anhydrous	7664-39-3	1000
Hydrogen Bromide	10035-10-6	5000
Hydrogen Chloride	7647-01-0	5000
Hydrogen Cyanide, Anhydrous	74-90-8	1000
Hydrogen Fluoride	7664-39-3	1000
Hydrogen Peroxide (52% by		

weight or greater)	7722-84-1	7500
Hydrogen Selenide	7783-07-5	150
Hydrogen Sulfide	7783-06-4	1500
Hydroxylamine	7803-49-8	2500
Iron, Pentacarbonyl	13463-40-6	250
Isopropylamine	75-31-0	5000
Ketene	463-51-4	100
Methacrylaldehyde	78-85-3	1000
Methacryloyl Chloride	920-46-7	150
Methacryloyloxyethyl Isocyanate	30674-80-7	100
Methyl Acrylonitrile	126-98-7	250
Methylamine, Anhydrous	74-89-5	1000
Methyl Bromide	74-83-9	2500
Methyl Chloride	74-87-3	15000
Methyl Chloroformate	79-22-1	500
Methyl Ethyl Ketone Peroxide (concentration greater than 60%)	1338-23-4	5000
Methyl Fluoroacetate	453-18-9	100
Methyl Fluorosulfate	421-20-5	100
Methyl Hydrazine	60-34-4	100
Methyl Iodide	74-88-4	7500
Methyl Isocyanate	624-83-9	250
Methyl Mercaptan	74-93-1	5000
Methyl Vinyl Ketone	79-84-4	100
Methyltrichlorosilane	75-79-6	500
Nickel Carbonly (Nickel Tetracarbonyl)	13463-39-3	150
Nitric Acid (94.5% by weight or greater)	7697-37-2	500
Nitric Oxide	10102-43-9	250
Nitroaniline (para Nitroaniline	100-01-6	5000
Nitromethane	75-52-5	2500
Nitrogen Dioxide	10102-44-0	250
Nitrogen Oxides (NO; NO(2); N2O4; N2O3)	10102-44-0	250
Nitrogen Tetroxide (also called Nitrogen Peroxide)	10544-72-6	250
Nitrogen Trifluoride	7783-54-2	5000
Nitrogen Trioxide	10544-73-7	250
Oleum (65% to 80% by weight; also called Fuming Sulfuric Acid)	8014-94-7	1000
Osmium Tetroxide	20816-12-0	100
Oxygen Difluoride (Fluorine Monoxide)	7783-41-7	100
Ozone	10028-15-6	100
Pentaborane	19624-22-7	100
Peracetic Acid (concentration greater 60% Acetic Acid; also called Peroxyacetic Acid)	79-21-0	1000
Perchloric Acid (concentration greater than 60% by weight)	7601-90-3	5000

Perchloromethyl Mercaptan	594-42-3	150
Perchloryl Fluoride	7616-94-6	5000
Peroxyacetic Acid (concentration greater than 60% Acetic Acid; also called Peracetic Acid)	79-21-0	1000
Phosgene (also called Carbonyl Chloride)	75-44-5	100
Phosphine (Hydrogen Phosphide)	7803-51-2	100
Phosphorus Oxychloride (also called Phosphoryl Chloride)	10025-87-3	1000
Phosphorus Trichloride	7719-12-2	1000
Phosphoryl Chloride (also called Phosphorus Oxychloride)	10025-87-3	1000
Propargyl Bromide	106-96-7	100
Propyl Nitrate	627-3-4	2500
Sarin	107-44-8	100
Selenium Hexafluoride	7783-79-1	1000
Stibine (Antimony Hydride)	7803-52-3	500
Sulfur Dioxide (liquid)	7446-09-5	1000
Sulfur Pentafluoride	5714-22-7	250
Sulfur Tetrafluoride	7783-60-0	250
Sulfur Trioxide (also called Sulfuric Anhydride)	7446-11-9	1000
Sulfuric Anhydride (also called Sulfur Trioxide)	7446-11-9	1000
Tellurium Hexafluoride	7783-80-4	250
Tetrafluoroethylene	116-14-3	5000
Tetrafluorohydrazine	10036-47-2	5000
Tetramethyl Lead	75-74-1	1000
Thionyl Chloride	7719-09-7	250
Trichloro (chloromethyl) Silane	1558-25-4	100
Trichloro (dichlorophenyl) Silane	27137-85-5	2500
Trichlorosilane	10025-78-2	5000
Trifluorochloroethylene	79-38-9	10000
Trimethyoxysilane	2487-90-3	1500

Footnote* Chemical Abstract Service Number

Footnote** Threshold Quantity in Pounds (Amount necessary to be covered by this standard.)

[57 FR 7847, Mar. 4, 1992]

◀ OSHA Regulations (Standards - 29 CFR) - Table of Contents

Index

About the Author

Bert Moniz is a materials engineering consultant with the DuPont Company, with over thirty years experience. He is involved in selecting materials of construction and in corrosion, failure analysis, and mechanical integrity programs across the company. Bert is co-editor of the reference book *Process Industries Corrosion* (National Association of Corrosion Engineers) and a textbook on metallurgy (American Technical Publishers). He is a registered professional engineer (P.Eng) in the Province of Ontario.